Water Quality - Factors and Impacts

Edited by Daniel Dunea

Published in London, United Kingdom

IntechOpen

Supporting open minds since 2005

Water Quality - Factors and Impacts
http://dx.doi.org/10.5772/intechopen.96701
Edited by Daniel Dunea

Contributors
Mohmadisa Hashim, Zahid Mat Said, Arijatul Wardah Ahmad, Nasir Nayan, Hanifah Mahat, Yazid Saleh, Koh Liew See, Thomas D. Shahady, Shadrack M. Kithiia, Benjamin Ezekeil Bwadi, Mohammed Bakoji Yusuf, Grace Audu, Ibrahim Abdullahi, Clement Yakubu Giwa, Jairo Lisboa Rodrigues, Mayra Santos, Vitor Silva, Thainá Andrade, Tamise Aquino, Bruno Batista, Bruna Freire, João Mesquita, Márcia Faria, Luiz Maia, Eduardo A. López-Maldonado, Mercedes T. Oropeza-Guzman, Carlos A. Quintero-González, Sampson Kwaku Agodzo, Maxwell Anim-Gyampo, Belinda Seyram Berdie, Enoch Bessah, Mario Alfonso Murillo-Tovar, Marilú Barrera-Olivarez, Josefina Vergara-Sánchez, María Luisa García Betancourt, Hugo Albeiro Saldarriaga Noreña, Francisco Martin-Romero, América Ramírez, Gabriela Eleonora Moeller Chávez, Satish Chandra Bhuyan, Subrat Kumar Bhuyan, Himanshu Bhusan Sahoo, Daniel Dunea

First published in London, United Kingdom, 2022 by IntechOpen
IntechOpen is the global imprint of INTECHOPEN LIMITED, registered in England and Wales, registration number: 11086078, 5 Princes Gate Court, London, SW7 2QJ, United Kingdom
Printed in Croatia

British Library Cataloguing-in-Publication Data
A catalogue record for this book is available from the British Library

Additional hard and PDF copies can be obtained from orders@intechopen.com

Water Quality - Factors and Impacts
Edited by Daniel Dunea
p. cm.
Print ISBN 978-1-83969-547-6
Online ISBN 978-1-83969-548-3
eBook (PDF) ISBN 978-1-83969-549-0

For EU product safety concerns:
IN TECH d.o.o., Prolaz Marije Krucifikse Kozulić 3, 51000 Rijeka, Croatia, info@intechopen.com or visit our website at intechopen.com.

Meet the editor

Daniel Dunea is a professor in the Department of Environmental Engineering, Valahia University of Targoviste, Romania. He received his Habilitation degree and Ph.D. from the University of Agronomical Sciences and Veterinary Medicine of Bucharest, Romania. He completed a Marie Curie Fellowship at Wageningen University & Research (WUR), Netherlands. In 2020, he was appointed a member of the prestigious National Council for Attestation of University Degrees, Diplomas and Certificates – Romanian Ministry of Education. He has published several books and more than fifty scientific articles indexed in the Web of Science (WoS). He has also reviewed manuscripts for more than thirty-five WoS journals. His main research interests are environmental monitoring and assessment and associated aspects.

Contents

Preface

The current knowledge about the water cycle at the river basin level is relatively incomplete due to random and complex processes that are difficult to measure or estimate in their integrity. Moreover, hydrological and water quality data for all rivers that make up a river network are scattered, discontinuous, or incomplete, which makes it difficult to assess river flow information for water resources management, flood risk, or pollutants' load assessment. Water quality is influenced by the characteristics of the river basin, such as land use and its geology, seasonal influence, river discharge, and chemical properties of tributary waters. Important factors in the quantitative balance of surface waters are precipitations and snowmelt with quantitative and qualitative influences, mainly due to the surface transport of germs, suspensions, pesticides, and fertilizers. Water quality parameters include a broad range of physicochemical and biological properties. Overall, there are six main indicators widely monitored to characterize the status of waters: water temperature, dissolved oxygen, pH, turbidity, nitrate chemicals, and bioindicators.

Consequently, despite the numerous approaches existing in the literature regarding water quality issues, there is a need to continuously monitor and assess the evolution of water resources and to find and implement new solutions for proper water management and protection. Thus, this book offers a wide image of the factors and impacts influencing water quality worldwide. It discusses the use of new analytical and monitoring methods and novel technologies for decontamination and protection of polluted waters.

The book is organized into three sections containing ten chapters.

The first section, "Monitoring and Assessment of Water Quality," contains four chapters that underline the need for monitoring water quality based on a series of relevant parameters and proper data analysis to avoid water shortages and food insecurity in the analyzed basins and areas by reducing the presence of contaminants.

The second section, "Factors of Water Quality Degradation," discusses the significant pressures and the impact of human activities on the state of surface waters and groundwater. The three chapters in this section discuss interesting aspects regarding the degradation and improvement of urban rivers, the effects of illegal artisanal gold mining operations, and the mobility of heavy metals in aquatic environments. Thus, a milestone is defined to start new research on excessive sediment and erosion mixed with microplastics, pathogens, toxic elements and compounds, and nutrients discharged in waters.

The third section, "Methods for Water Quality Improvement," contains three chapters that point out the implications of proper decontamination methods for various polluted waters based on bio-polyelectrolytes, natural products including agricultural waste, and other waste materials.

Water quality is a key environmental concern at the global level, having a significant role in promoting socioeconomic development but with the need of maintaining viable and functional ecosystems. The protection of water quality requires improved monitoring and assessment methods included in reliable watershed management plans.

I would like to thank Dr. Petre Bretcan for the useful discussions and recommendations regarding various aspects related to water quality presented in this book. I am also grateful to Sara Debeuc and Anja Filipovic at IntechOpen for their professional support throughout the publication process.

Daniel Dunea
Department of Environmental Engineering,
Valahia University of Târgoviște,
Târgoviște, Romania

Section 1

Monitoring and Assessment of Water Quality

Chapter 1

Water Quality and Anthropogenic Pressures in a Changing Environment: The Arges River Basin, Romania

Daniel Dunea

Abstract

The objective of this work was to present several benchmarks regarding the water quality at hydrological basin level under increasing anthropogenic pressures. The first part briefly describes the sources of water pollution, the hydromorphological pressures, and the main water quality parameters widely used for the assessment. The second part presents as an example the dynamics of several water quality parameters recorded between 2007 and 2014 downstream of Argeș River, Romania, near the confluence with the Danube River. Argeș River supplies water for several important Romanian cities including Bucharest, and from here comes the rationale of the work, which envisages characterizing water quality status to substantiate proper water management. The following parameters were statistically analyzed: water temperature, suspended solids, pH, dissolved oxygen, biochemical oxygen demand, ammonium, nitrates, nitrites, and dissolved heavy metals. The factor analysis results showed that the first factor contains temperature and dissolved oxygen, the second has the heavy metals, the third groups have the ammonium and pH, the fourth contains the TSS and nitrites, while the fifth is formed by BOD5 and nitrates. Water quality plays a significant role in promoting socioeconomic development and maintaining viable ecosystems. The protection of water quality requires improved monitoring and reliable watershed management plans.

Keywords: EU Water Framework Directive, pollution sources, Argeș River, water monitoring, biological oxygen demand, heavy metals

1. Introduction

1.1 The current state of knowledge about water quality protection in the European Union

The EU Water Framework Directive (WFD), adopted on 23 October 2000 and implemented since 22 December 2000, when it was published in the Official Journal of the European Union (OJ L 327), has as its main objective the achievement of good status for all bodies of water (surface and groundwater), and for artificial bodies, the good ecological potential was defined [1]. This requires an analysis of anthropogenic pressures and their impact, as set out in Article 5 of the WFD, which states: "*Each Member State must review the impact of human activities on the state of surface and groundwater in each district of the river basin or for a portion of a district of an international river basin situated in its territory.*" WISE is the Water Information System for Europe, an information gateway for stakeholders regarding the European water issues comprising data collected by dedicated EU institutions [2].

Watershed management plans are management tools in Integrated Water Resources Management, which generally contain descriptions of water resources in a drainage basin and the associated management schemes and applications considering the economic efficiency and the social component in the use of water as important pillars [3]. River basin management plans come as a requirement of the WFD and are means of achieving the protection, improvement, and sustainable use of water resources. This includes fresh surface water (lakes, rivers, and streams), groundwater, and ecosystems such as some wetlands that depend on groundwater, estuaries, and coastal waters [4]. At the river basin level, it is important to identify the pressures and estimate the anthropogenic impact on the state of surface and groundwater, this aspect being achieved by identifying the activities carried out at the river basin level and significant pressures, followed by the impact assessment and measures [5]. Consequently, the actions envisage the improvement of the water body conditions [6].

The identification of significant pressures as well as the impact of human activities on the state of surface waters and groundwater [7] takes into account the following aspects:

- point sources of significant pollution;
- sources of diffuse pollution, including land use impact;
- hydromorphological pressures.

Water resources management is the activity of planning, development, distribution, and optimal management of water resources, water resources management planning takes into account all competing water demands and tries to allocate water to meet all uses and requirements. Within the management of water resources, the following major categories of problems have been identified: pollution with organic substances, pollution with nutrients, pollution with priority hazardous substances, and hydromorphological pressures [8].

Pollution with organic substances is due to wastewater discharges from point and diffuse sources represented by human settlements, industrial and agricultural sources having a significant impact on aquatic ecosystems [9]. Nutrient pollution of water refers to contamination with excessive nutrient inputs, this being the main cause of eutrophication of surface waters, in which excess nutrients, usually nitrogen or phosphorus, stimulate the growth of algae. Sources of nutrient pollution include spills from agricultural fields and pastures, discharges from septic tanks, and emissions from combustion [10]. Pollution with hazardous substances emitted into water bodies occurs both directly and indirectly through a series of diffuse or point sources, these substances are toxic and have persistence and bioaccumulation in the aquatic environment [11].

The objective of this work was to establish the key benchmarks regarding water quality resilience under increasing anthropogenic pressures in a changing environment determined by climate change. In the first part, the sources of water pollution (point and diffuse), the hydromorphological pressures, and the main water quality parameters have been briefly described. In the second part, the dynamics of several water quality parameters recorded downstream of Argeș River, Romania, near the confluence with the Danube

River, from 2007 to 2014 were presented as an example for water quality dynamics in conjunction with the anthropogenic impact and enforced environmental regulations.

1.2 Significant point sources of water pollution

According to the WFD, there are certain limits above which pressures become significant, so the size of the pressure is compared to a relevant limit value for a body of water [1]. Regarding the discharge of treated or untreated water into surface waters, the following significant pressures are identified:

- human agglomerations with more than 2000 equivalent inhabitants and wastewater collection systems, with or without wastewater treatment plants, but also agglomerations of less than 2000 equivalent inhabitants with centralized or unitary sewerage systems are considered point sources of significant pollution [1].

- an industry with installations covered by the Directive on integrated pollution prevention and control; with units that discharge hazardous substances beyond the limits of the legislation in force, or other units that do not comply with the legislation in force and evacuate in surface water resources [1].

- agriculture with livestock farms covered by the Directive on integrated pollution prevention and control—96/61/EC (IPPC Directive); with farms that discharge hazardous substances beyond the limits of the legislation and do not comply with it [12].

Human agglomerations contribute with significant amounts of organic matter (CCO and BOD_5), nutrients (total nitrogen and total phosphorus), and heavy metals (Cu, Zn, Cd, Ni, Pb, Cr, Hg) when it comes to the discharge of pollutants into surface water resources, as there are still numerous human settlements that do not comply with the requirements of the Urban Wastewater Treatment Directive (Directive 91/271/EEC) [13]. There are specific sources of industrial and agricultural pollution that must meet certain requirements, such as integrated pollution prevention and control, pollution from hazardous substances, protection of waters against nitrate pollution from agricultural sources, and pollution occurring from major accidents [14].

1.3 Significant diffuse sources of pollution including land use

Water pollution, such as sewage or industrial effluent pollution, is normally easier to monitor because it generally occurs from a single source, but diffuse water pollution results from multiple sources [15]. Diffuse water pollution is a widespread problem and it is important to know the extent to which different sources of diffuse pollution have an impact on water quality [16], agriculture being one of the main sources of diffuse pollution. Various activities contribute to diffuse pollution indicating agriculture, forestry, mining, construction, and urban life, and the local climate, geology, and other natural phenomena that can influence the size and extent of the problem. In agriculture, diffuse pollutants include sludge from soil erosion, nutrients from fertilizer application, or non-compliance with pesticide handling and chemical application legislation—Directive 2009/128/EC [17].

As in the case of point sources of significant pollution, the main categories are represented by:

- human agglomerations without wastewater collection systems and compliant landfills;

- agriculture, there are zootechnical farms that do not have systems for storage or use of compliant manure and as a result, there are vulnerable areas in terms of pollution with nitrates from agricultural sources, and by non-compliance with current legislation frequently appear areas polluted with pesticides;

- regarding the industry, the diffuse sources of pollution are represented by the storage of waste in non—compliant landfills, raw material landfills, accidental pollution within the units, the existence of abandoned industrial sites.

Eutrophication is the enrichment of water with nutrients such as nitrate or phosphate, causing an accelerated development of algae and higher plants, which leads to unwanted disturbance of the balance of organisms in the water but also to water quality in general [18]. Thus, habitat disturbances occur for fish and invertebrate species, the development of toxic algae with an impact on the fishing industry, and the flourishing of recreational waters lead to the closure of the navigation or recreational use area with an impact on the tourism and leisure industry [19]. Loss of nutrients or agrochemicals in soil and water in addition to affecting the environment is also an agricultural financial loss. In addition, groundwater is endangered due to leakage or percolation of nutrients and pesticides from the land surface. Diffuse agricultural pollution is mainly associated with soil particles, pesticides, and other potentially toxic chemicals, including veterinary medicines, nutrients, pathogens, bacteria from animal waste, and manure spread on the ground [20].

1.4 Significant hydromorphological pressures

The main hydromorphological pressures identified in the risk analyzes are hydropower, flood protection, navigation, and agriculture. There are other activities of some importance, such as urbanization, water use for irrigation, outdoor recreational activities, and fishing. Hydromorphological pressures are often generated by the performance of hydrotechnical works, such as dams for hydropower generation, flood protection, as well as water capture or construction of navigation canals [21].

Hydromorphological pressures affect watercourses in river basins, the most important hydromorphological pressures being caused by:

- regularization and dam works that bring changes related to the morphology of watercourses;

- execution of derivations with the role of supplementing the tributary flow for certain accumulations and ensuring the water requirement for the afferent localities;

- storage lakes that are built for drinking and industrial water supply, having a role of protection against floods, or with those used for irrigation, fish farming, or for energy purposes;

- navigable canals, that modify the morphology of the riverbed, the navigation representing a significant pressure on the waters due to the risk of accidental pollution.

These hydromorphological changes such as dams, reservoirs, canals, diversions with changes in the profile of the surface water body, can in turn cause changes such as interruptions of sediment transport, changes in hydraulic and hydrological characteristics (reduced water flow), loss of areas flooding or drying of wetlands, as well as the direct damage to the biota and disruption of biological continuity. The impact of these changes and the subsequent effects lead to the disappearance of aquatic communities [22].

1.5 Surface water quality categories

In general, the quality of surface water is determined by its loading with mineral or organic substances, suspended particles, living organisms, and dissolved gases, and in terms of water quality characteristics, the following terms are generally used [23]:

- water quality indicators;
- water quality parameters;
- water quality criteria;
- standardized water quality values.

Water quality indicators provide basic information and help identify the trends of the changes in water quality over time. The most important indicators considered in the monitoring strategies and plants are dissolved oxygen, biological oxygen demand, water temperature, pH, nitrate (NO_3), nitrite (NO_2), and ammonia (NH_3), metals, transparency, turbidity, coliforms, etc. [24, 25].

Dissolved oxygen (DO) is essential for plants and animals, but at high levels, in water, it can be harmful to fish and other aquatic organisms. Dissolved oxygen is measured in milligrams per liter (mg/L). Expected levels range between 4.0 and 12.0 mg/L [24].

Biological oxygen demand (BOD) is the amount of oxygen consumed by bacteria in the breakdown of organic matter. It also includes the oxygen needed to oxidize various chemicals in water, such as ammonia. BOD is determined by measuring the level of dissolved oxygen in a freshly collected sample and comparing it with the level of dissolved oxygen in a sample that was collected at the same time but incubated under specific conditions for several days. The difference in oxygen reading between the two samples is recorded in units of mg/L. Unpolluted, natural water should have a BOD level of 5 mg/L or even less, and wastewater may have BOD levels varying from 150 to 300 mg/L [25].

Water temperature affects many other parameters of water, including the amount of available dissolved oxygen, the types of plants and animals present, and the susceptibility of organisms to parasites, pollution, and diseases. The temperature is measured in degrees Celsius (°C). Seasonal trends: May–October: 22–35°C, from November to April: 2–27°C [24].

The *pH* test measures the alkalinity or acidity of the water. A pH of 7 is neutral, below 7 is acidic, and above 7 it is basic or alkaline. Acid rain, car leaks, or coal-fired power plants cause a drop in water pH [24]. Pollution from accidental spills, agricultural spills, and sewage spills can also change the pH. While juvenile fish and insect larvae are sensitive to low pH (acid), extreme values at each end of the scale can be lethal to most organisms. Expected levels are 6.5–9.0 [25].

Nitrogen from the atmosphere or soil can undergo many complex chemical and biological changes, being a necessary nutrient for the growth of all living organisms. Nitrogen is found in natural waters in various forms: *nitrate* (NO_3), *nitrite* (NO_2), and *ammonia* (NH_3). The test results are usually expressed as nitrate-nitrogen. Ammonia is the least stable form of nitrogen and thus it is difficult to measure it accurately. In large quantities, nitrates lead to excess growth of algae. Under certain conditions, high levels of nitrates (10 mg/L or more) in drinking water can be toxic to humans and have also been linked to serious illness and even death in infants. Nitrates are expressed in milligrams per liter (mg/L) [25]. *Ammonia* is one of the most important pollutants in the aquatic environment due to its extremely toxic nature. It is discharged in large quantities into industrial, municipal, and agricultural wastewater [24]. *Nitrite* (NO_2) is extremely toxic to the aquatic environment but is usually present in very small amounts in most natural water systems because it is rapidly oxidized to nitrate. Organic nitrogen and ammonia can be determined together and have been referred to as "Kjeldahl nitrogen, or TKN", a term that reflects the technique used in their determination (edition 19, Standard Methods, 1995).

Transparency is the quality of water to let the light energy through. Sunlight provides photosynthetic energy and determines the depth at which algae and other plants can grow. A change in the clarity of the water can be observed after heavy rains when mud and other debris cause decreased visibility [24].

The *turbidity* of water represents its lack of transparency being caused by suspended solid particles and plankton (microscopic plants and animals), which cannot be individualized with the naked eye. A low level of turbidity indicates a healthy ecosystem with moderate amounts of plankton, but a higher level of turbidity raises many problems for flow systems, surface water temperature can rise above normal, as particles suspended close to the water surface facilitate absorption heat from the sun. Suspended particles can carry nutrients, pesticides, and other pollutants and cloudy waters may have low levels of dissolved oxygen [25].

Water quality standards are based exclusively on scientific data on the relationship between pollutant concentrations and their effects on the environment and human health. The quality criteria provide a framework for the control of discharges and pollutant emissions [23]. Water quality standards define the water quality objectives in a body of water by designating beneficial uses and establishing criteria to protect those uses [14]. Beneficial uses may include fishing, swimming, aquatic habitat, navigation, agriculture, etc.

2. Case study: the Argeș River, Romania

2.1 Study area

The water quality parameters were recorded within the Transnational Monitoring Network of the Danube River (TNMN) in Argeș River located in Southern Romania. Argeș River is a left tributary of the Danube being 350 km long and having a basin area of approximately 12,550 km^2, which is 5.3% of Romania's total area (**Figure 1**). The source is located in the Făgăraș Mountains, and it flows into the Danube River at Oltenița.

2.2 Methodology

Corine Land Cover [26] database was used to perform the land use/land cover analysis using the ArcGIS Desktop functions (https://www.ecologic.eu/sites/default/files/project/2021/Land%20Use%20Analysis%20final%20en.pdf).

The water quality was monitored on Argeș River near the confluence with the Danube River at Clătești village (**Figure 2**) within the TNMN, International Commission for the Protection of the Danube River (ICPDR)—https://www.icpdr.org/wq-db/. From the multitude of the parameters collected in the dataset between

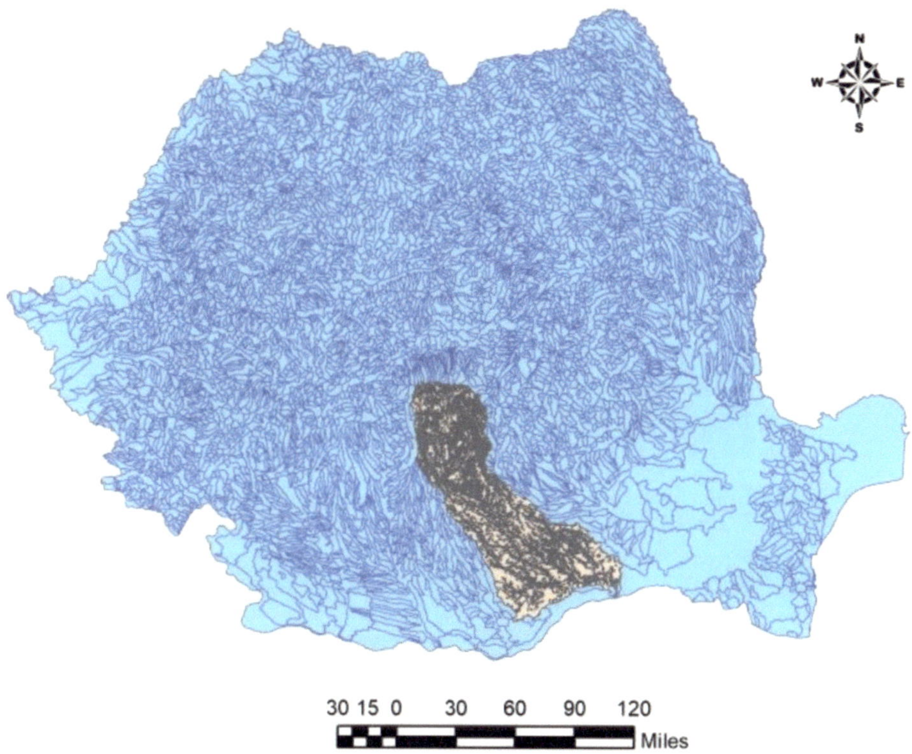

Figure 1.
Argeş River hydrological basin (Code X.1) position in Romania (blue lines represent the delimitations of the subbasins).

2007 and 2014, the following were considered for presentation in this work (n = 95 for each parameter): water temperature (°C), suspended solids—TSS (mg/L), pH, dissolved oxygen—DO (mg/L), biochemical oxygen demand—BOD_5 (mg/L), nitrogen nutrients i.e., ammonium (NH_4-N) (mg/L), nitrates (mg/L), and nitrites (mg/L) and heavy metals (μg/L) i.e., dissolved nickel, chromium, and lead. For these parameters, the sampling, storage, and handling of samples were performed with a monthly frequency being carried out according to the current standards [14]. The start of the time series was considered in the year 2007 because Romania became a member of the EU in that year, and consequently the water quality standards needed to be in agreement with EU legislation.

Statistical analysis was performed using descriptive and associative statistics using the SPSS software (SPSS Inc., Chicago, IL, 2011). Factor analysis considered principal component analysis (PCA) based on Varimax with Kaiser normalization to reduce the number of factors that explain the variability in the dataset.

2.3 Results

In order to have an image of the potential impact of land use within the surfaces existing in the basin of Arges river that drains the waters, the land use analysis was applied (**Figure 3**). The main cities supplied with water in the Arges river basin include the capital Bucharest, and other important cities e.g., Pitești, Curtea de Argeș, Câmpulung, Găești.

Oltenița. The category 112—discontinuous urban fabric occupies 4%. The 211—non-irrigated arable land category reaches 46% showing that the area is important for agriculture. Furthermore, the basin contains a key forest resource

Figure 2.
Position of the monitoring point at Clătești village on Argeș River (arrow) (44.14500N, 26.59900E) https://geoportal.ancpi.ro/portal/apps/webappviewer/index.html?id=5fca89129f2f466882bb7c64e6fd3d98#.

(approximately 3283 km^2) which represents 26.2% of the river basin area and 5.2% of the national forest fund. Watercourses (511 class) represent 12% because the Argeș River gathers 178 codified watercourses with a length of 4579 km (5.8% of the total length of codified watercourses in Romania). The density of watercourses is 0.36 km/km^2.

Figure 4 shows the overall results of the land use analysis.

Land use may have a direct influence on the water quality in the watercourses because of direct discharge and runoff.

DOI: http://dx.doi.org/10.5772/intechopen.101790

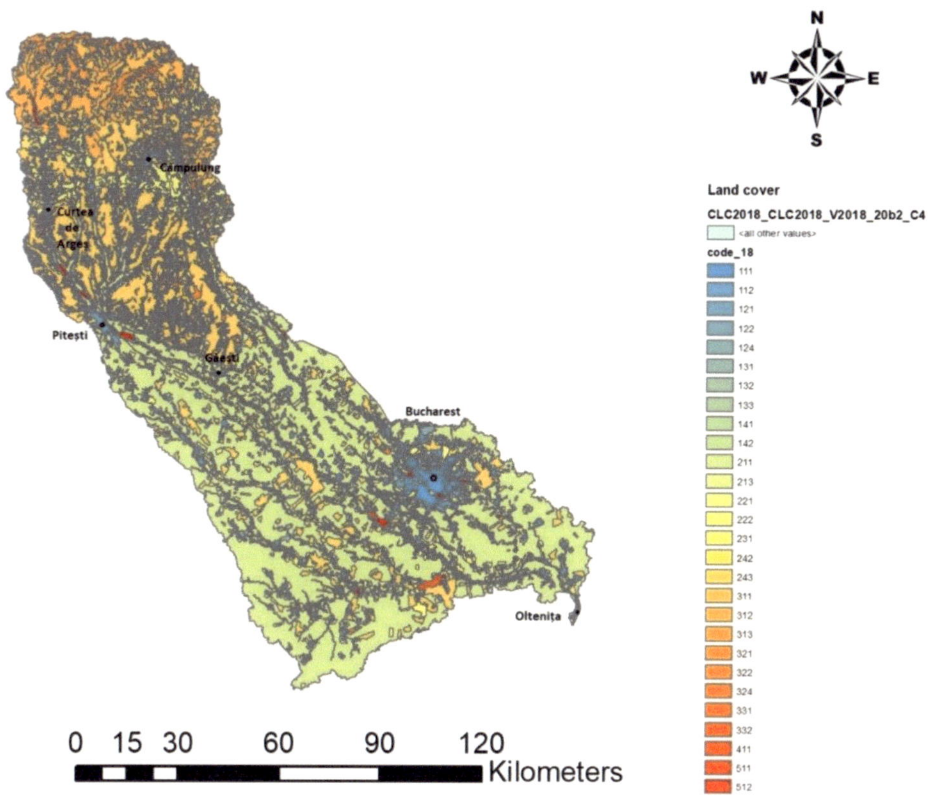

Figure 3.
Land use in the Argeș hydrological basin according to Corine Land Cover.

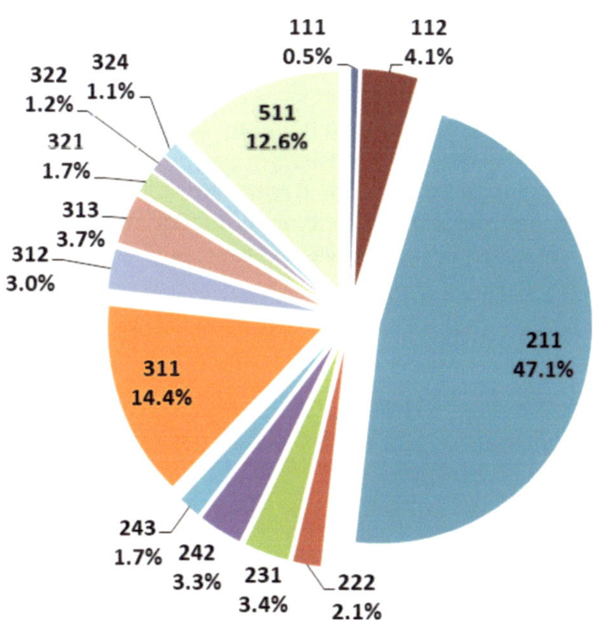

Figure 4.
Analysis of the land use in the Argeș hydrological basin (predominant classes: 112—discontinuous urban fabric 4%; 211—non-irrigated arable land 47%; 311—broad-leaved forest; 14%; 511—watercourses 12%)—details: https://land.copernicus.eu/user-corner/technical-library/corine-land-cover-nomenclature-guidelines/html.

The time series recorded between 2007 and 2014 were plotted for the envisaged parameters to show the dynamics. **Figure 5** presents the fluctuations of pH values for the water samples collected each month. The associated trend line shows a pH value around 7.5, ranging from 6.53 to 8.26 (**Table 1**).

Figure 6 shows the fluctuation of water quality parameters related to oxygen, nutrients, and heavy metals during the monitored period. BOD_5 showed an increasing trend, while DO remained almost constant. Nitrogen parameters decreased from 2007 to 2014. The main statistics for each parameter are presented in **Table 1**.

Dissolved Ni and Pb showed the highest variations in the dataset and these indicators show the magnitude of the anthropogenic impact on the water quality due to the discharge of industrial wastewaters and atmospheric deposition. **Table 2** presents the correlation matrix resulting from applying the Pearson method. The strongest inverse correlations ($p < 0.01$) were found between BOD_5 and dissolved lead, TSS and nitrites, temperature and nitrates, pH and temperature, temperature, and dissolved oxygen. Positive correlations ($p < 0.01$) were established between the dissolved heavy metals suggesting that their concentrations rise together.

The application of factor analysis to the dataset (a matrix of 95 objects by 11 variables) was made using the Varimax with Kaiser normalization [27] to reduce the number of factors that explains the variability. Five components were extracted and from the rotated matrix the factors were organized based on the relevant factor loadings (>0.6). The results showed that the first factor contains temperature and dissolved oxygen, the second has the heavy metals, the third groups the ammonium and pH, the fourth contains the TSS and nitrites, while the fifth is formed by BOD_5 and nitrates. These factors accounted for a cumulative variance of 72.3% of the total variability in the dataset.

Figure 7 summarizes the results of the factor analysis applied to the water quality dataset recorded at Clătești near the confluence of Argeș River with the Danube River.

When compared to the Romanian NTPA-013/2002 limit values (surface waters used for drinking source), the reported concentrations presented exceeding for dissolved Ni, Cr, and Pb. The presence of heavy metals at Clătești monitoring point on Argeș River is a clear indicator of the pressure on the water quality. Furthermore, ammonium exceeded the limit values for class A3 (lowest) and BOD_5 for class A2, respectively. These suggest an increased pollutant load near the discharge of Argeș in the Danube River.

Within the hydrographic basin of the Argeș river, it is necessary to identify and quantify the significant pressures either from punctual sources, with discharges of treated waters or untreated in surface waters (sources of urban pollution/human

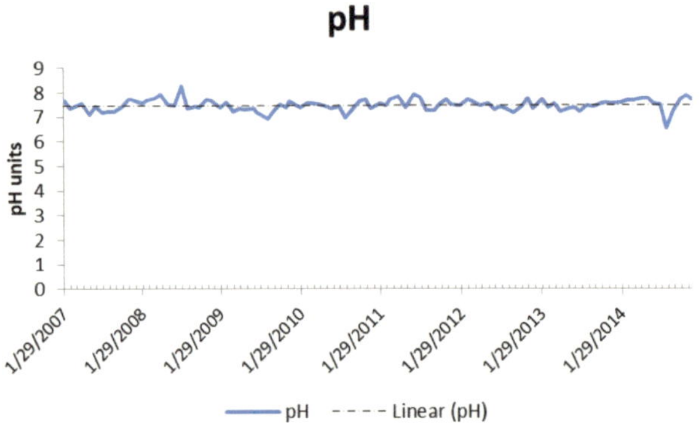

Figure 5.
pH time-series recorded between 2007 and 2014 (monthly concentrations).

Parameter	BOD_5	Suspended solids	Ammonium	Nitrates	pH	Water temperature	DO	Nitrites	Dissolved Ni	Dissolved Cr	Dissolved Pb
Units	mg/L	mg/L	mg/L	mg/L	—	°C	mg/L	mg/L	µg/L	µg/L	µg/L
Count	95	95	95	95	95	95	95	95	95	95	95
Average	4.45	64.09	2.93	1.27	7.48	14.92	7.60	0.07	3.95	1.31	0.69
Median	4.20	54.00	2.65	1.08	7.48	15.00	7.16	0.07	2.40	1.00	0.45
Minimum	2.51	20.00	0.33	0.01	6.53	1.50	4.96	0.01	1.00	0.50	0.16
Maximum	7.20	214.00	7.79	10.70	8.26	29.50	12.59	0.20	49.85	6.80	5.09
Standard deviation	1.07	37.07	1.69	1.19	0.24	8.48	1.85	0.04	6.17	0.78	0.74
Coefficient of variation (%)	24.0	57.8	57.7	94.0	3.2	56.8	24.3	50.6	156.4	59.5	106.8
Skewness	0.58	2.19	0.76	5.42	−0.36	−0.04	0.66	0.90	5.34	4.04	4.24
Kurtosis	−0.34	5.44	0.36	41.91	2.48	−1.34	−0.39	1.21	34.2	25.52	21.27

Table 1.
Descriptive statistics of the parameters recorded at Clătești monitoring point on Argeș River between 2007 and 2014.

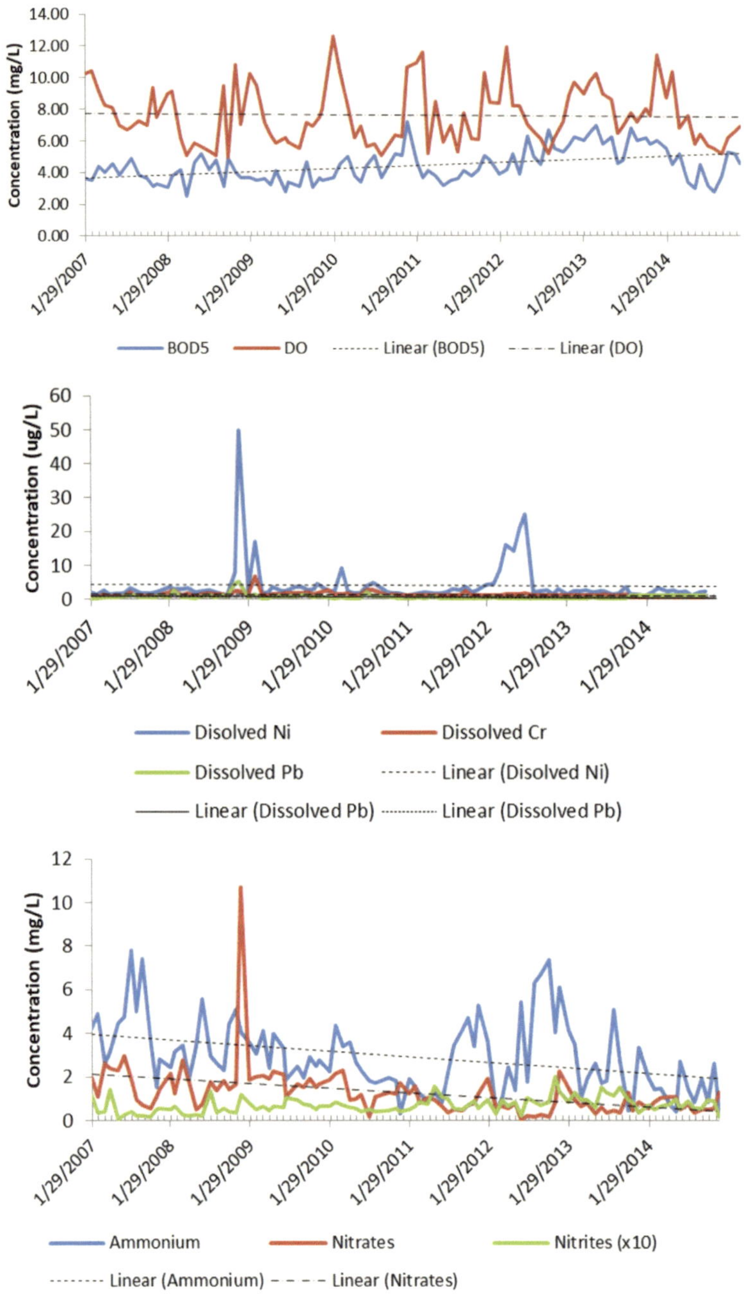

Figure 6.
Dynamics of oxygen-related parameters, dissolved heavy metals, and nitrogen nutrients in the monitoring point on Argeș River.

settlements and/or industrial sources with wastewaters discharged into streams due to non-compliance with the maximum permitted concentrations [28]), or diffuse pollution, from agriculture and other sources.

An investigative monitoring program is needed to identify the causes of exceeding the limits provided in quality standards and other regulations in the field of water management, to establish the causes for which a body of water in the Argeș basin cannot achieve the established environmental objectives, but also to quantify the impact on water quality, providing information on the programs of measures

DOI: http://dx.doi.org/10.5772/intechopen.101790

	BOD_5	TSS	Ammonium	Nitrates	pH	Temp	DO	Nitrites	Diss. Ni	Diss. Cr	Diss. Pb
BOD_5	1.00	−0.04	0.12	**−0.24**	0.01	−0.07	0.18	**0.20**	0.01	−0.15	**−0.27**
	—	*0.68*	*0.24*	*0.02*	*0.92*	*0.51*	*0.08*	*0.05*	*0.89*	*0.16*	*0.01*
TSS	—	1.00	−0.13	0.07	0.06	−0.12	**0.21**	**−0.27**	−0.10	−0.13	−0.04
	—	—	*0.20*	*0.50*	*0.54*	*0.23*	*0.04*	*0.01*	*0.35*	*0.22*	*0.69*
Ammonium	—	—	1.00	0.12	**−0.24**	0.08	−0.03	−0.13	0.18	0.14	−0.07
	—	—	—	*0.25*	*0.02*	*0.43*	*0.81*	*0.21*	*0.08*	*0.19*	*0.52*
Nitrates	—	—	—	1.00	0.10	**−0.26**	0.10	0.03	−0.06	0.12	0.12
	—	—	—	—	*0.33*	*0.01*	*0.32*	*0.80*	*0.55*	*0.26*	*0.25*
pH	—	—	—	—	1.00	**−0.36**	0.15	0.18	−0.08	−0.07	0.09
	—	—	—	—	—	*0.00*	*0.14*	*0.08*	*0.43*	*0.51*	*0.41*
Temp	—	—	—	—	—	1.00	**−0.76**	0.05	0.02	−0.06	**−0.22**
	—	—	—	—	—	—	*0.00*	*0.61*	*0.88*	*0.54*	*0.04*
DO	—	—	—	—	—	—	1.00	0.01	−0.02	−0.05	0.05
	—	—	—	—	—	—	—	*0.94*	*0.82*	*0.66*	*0.65*
Nitrites	—	—	—	—	—	—	—	1.00	−0.08	**−0.22**	**−0.20**
	—	—	—	—	—	—	—	—	*0.42*	*0.03*	*0.05*
Diss. Ni	—	—	—	—	—	—	—	—	1.00	**0.38**	**0.50**
	—	—	—	—	—	—	—	—	—	*0.00*	*0.00*
Diss. Cr	—	—	—	—	—	—	—	—	—	1.00	**0.34**
	—	—	—	—	—	—	—	—	—	—	*0.00*
Diss. Pb	—	—	—	—	—	—	—	—	—	—	1.00
	—	—	—	—	—	—	—	—	—	—	—

Calculated p is italicized; significant correlations are bolded.

Table 2.
Correlation matrix of the parameters monitored on Argeș River at Clătești between 2007 and 2014.

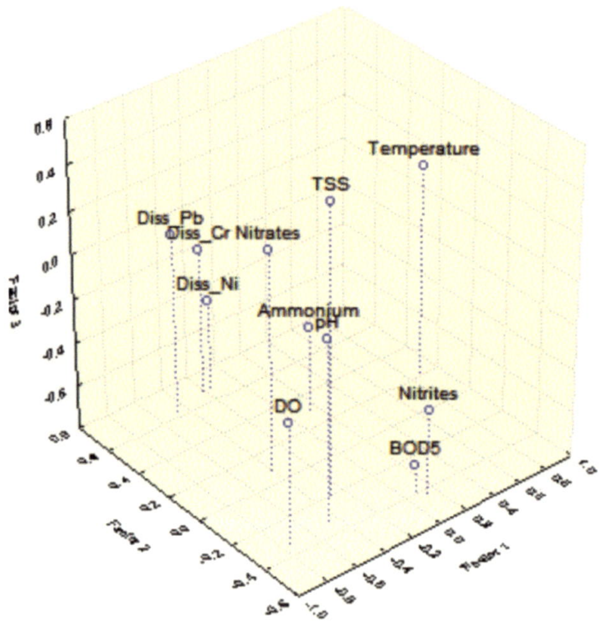

Figure 7.
The component plot in rotated space for the water quality parameters (1st factor, temperature, and dissolved oxygen, 2nd factor, the heavy metals, the 3rd factor, ammonium and pH, 4th factor, TSS and nitrites; 5th factor: BOD_5 and nitrates).

needed to achieve the environmental objectives and the specific measures needed to remediate the effects of accidental pollution [28].

Such tools and measures lead to the completion of knowledge on water quality, to the achievement of an optimal qualitative assessment, to the testing of hypotheses on the assessment of pressures and impact [29].

The aim is also to conserve water-dependent habitats and species, to enable the sustainable use of resources and the efficient management of water resources. These measures are necessary to reduce the hydromorphological pressures, the effects of climate change, the phenomenon of eutrophication in water bodies as well as the protection of groundwaters. At the same time, a better understanding of the situation regarding the spatial distribution of habitats and species is needed in line with the trends in land-use change.

3. Conclusions

Water quality has emerged as one of the main concerns around the world recently, as it plays a significant role in promoting socio-economic development as well as maintaining viable ecosystems. The impact of water quality degradation has serious consequences, including eutrophication, sedimentation, harmful algae proliferation, and hypoxia, which has a negative impact on human settlements and ecosystems in terms of health and economy.

The Argeș River from Romania encounters significant pressures from anthropogenic factors as well as other rivers [30] that pass highly inhabited areas with moderate efficiency of wastewater treatment. In addition to the potentially significant pressures already presented above, other types of activities/pressures may affect the condition of water bodies such as accidental pollution, fishing, and aquaculture activities, ballast and sand extraction from minor riverbeds, forest exploitation, unidentified pressures, etc. [31].

The following recommendations are suggested to maintain or improve future water quality at the hydrological basin level:

a. stopping the discharge of untreated wastewater directly into water bodies and the implementation of modern wastewater treatment facilities for the proper treatment of wastewaters before discharging. Therefore, strict environmental regulations are required to manage these negative aspects;

b. the use of state-of-the-art technologies of water management, and monitoring to identify the causes and sources of pollution, as well as improvement of policy strategies;

c. capacity building and environmental ethics through communication, environmental education, training, and awareness that will improve waste management and water use;

d. cooperating with local stakeholders of water use, such as farmers, local institutions and non-governmental organizations, to facilitate the acceptance of new technologies and environmental policies in the context of climate change.

Acknowledgements

This work was performed in the framework of a grant of the Romanian National Authority for Scientific Research, CNDI– UEFISCDI, project number PN-III-1.2-PCCDI-2017-0721, https://inter-aspa.ro/.

Conflict of interest

The author declares no conflict of interest.

Thanks

The author wants to thank the Editorial team of INTECH and the team of the INTER-ASPA project for their support in finalizing this chapter.

Author details

Daniel Dunea
Valahia University of Târgoviște, Târgoviște, Romania

*Address all correspondence to: dan.dunea@valahia.ro

References

[1] The EU Water Framework Directive—Integrated river basin management for Europe [Internet]. Available from: https://ec.europa.eu/environment/water/water-framework/index_en.html [Accessed: July 20, 2021]

[2] The Water Information System for Europe—WISE [Internet]. Available from: https://water.europa.eu/freshwater [Accessed: July 20, 2021]

[3] Momblanch A, Pedro-Monzonís M, Solera A, Andreu J. Chapter 3—Water accounting for integrated water resources management: Experiences and recommendations. In: Friesen J, Rodríguez-Sinobas L, editors. Advances in Chemical Pollution, Environmental Management and Protection. Vol. 3. Amsterdam: Elsevier; 2018. pp. 63-96. DOI: 10.1016/bs.apmp.2018.08.001

[4] Grizzetti B, Liquete C, Pistocchi A, Vigiak O, Zulian G, Bouraoui F, et al. Relationship between ecological condition and ecosystem services in European rivers, lakes and coastal waters. Science of the Total Environment. 2019;**671**:452-465. DOI: 10.1016/j.scitotenv.2019.03.155

[5] Moore S. Toward effective river basin management (RBM): The politics of cooperation, sustainability, and collaboration in the Delaware River basin. Journal of Environmental Management. 2021;**298**:113421. DOI: 10.1016/j.jenvman.2021.113421

[6] Santos JI, Vidal T, Gonçalves FJM, Castro BB, Pereira JL. Challenges to water quality assessment in Europe—Is there scope for improvement of the current water framework directive bioassessment scheme in rivers? Ecological Indicators. 2021;**121**:107030. DOI: 10.1016/j.ecolind.2020.107030

[7] Mishra A, Alnahit A, Campbell B. Impact of land uses, drought, flood, wildfire, and cascading events on water quality and microbial communities: A review and analysis. Journal of Hydrology. 2021;**596**:125707. DOI: 10.1016/j.jhydrol.2020.125707

[8] Water Resources Management [Internet]. Available from: https://www.worldbank.org/en/topic/waterresourcesmanagement [Accessed: July 20, 2021]

[9] National Service Center for Environmental Publications (NSCEP). Control of organic substances in water and wastewater [Internet]. Available from: https://nepis.epa.gov/Exe/ZyPURL.cgi?Dockey=20007S8Y.TXT [Accessed: July 20, 2021]

[10] Caruso BS, O'Sullivan AD, Faulkner S, et al. Agricultural diffuse nutrient pollution transport in a mountain wetland complex. Water, Air, and Soil Pollution. 2013;**224**:1695. DOI: 10.1007/s11270-013-1695-x

[11] DeForest DK, Brix KV, Adams WJ. Assessing metal bioaccumulation in aquatic environments: The inverse relationship between bioaccumulation factors, trophic transfer factors and exposure concentration. Aquatic Toxicology. 2007;**84**(2):236-246. DOI: 10.1016/j.aquatox.2007.02.022

[12] Council Directive 96/61/EC concerning integrated pollution prevention and control [Internet]. Available from: http://www.fao.org/faolex/results/details/en/c/LEX-FAOC034860/ [Accessed: July 20, 2021]

[13] Urban Wastewater Treatment Directive (Directive 91/271/EEC) [Internet]. Available from: https://ec.europa.eu/environment/water/water-urbanwaste/legislation/directive_en.htm [Accessed: July 20, 2021]

[14] Dunea D, Bretcan P, Tanislav D, Serban G, Teodorescu R, Iordache S,

et al. Evaluation of water quality in Ialomita River Basin in relationship with land cover patterns. Water. 2020;**12**(3):735. DOI: 10.3390/w12030735

[15] Dunea D, Iordache Ş. Time series analysis of the heavy metals loaded wastewaters resulted from chromium electroplating process. Environmental Engineering and Management Journal (EEMJ). 2011;**10**:469-482

[16] Foster S, Chilton J. Policy experience with groundwater protection from diffuse pollution—A review. Current Opinion in Environmental Science and Health. 2021;**23**:100288. DOI: 10.1016/j.coesh.2021.100288

[17] Directive 2009/128/EC of the European Parliament and of the Council of 21 October 2009 establishing a framework for Community action to achieve the sustainable use of pesticides [Internet]. Available from: https://eur-lex.europa.eu/legal-content/EN/ALL/?uri=CELEX%3A32009L0128 [Accessed: July 20, 2021]

[18] Chislock MF, Doster E, Zitomer RA, Wilson AE. Eutrophication: Causes, Consequences, and Controls in Aquatic Ecosystems. Nature Education Knowledge. 2013;**4**(4):10

[19] Morris JG. Harmful algal blooms: An emerging public health problem with possible links to human stress on the environment. Annual Review of Energy and the Environment. 1999;**24**:367-390

[20] Schindler DW. Recent advances in the understanding and management of eutrophication. Limnology and Oceanography. 2006;**51**:356-363

[21] Fazelpoor K, Martínez-Fernández V, García de Jalón D. Exploring the hydromorphological response to human pressure in Tagus River (1946-2014) by complementary diagnosis. Catena. 2021;**198**:105052. DOI: 10.1016/j.catena.2020.105052

[22] Radu O, Cimpeanu SM, Teodorescu RI, Bucur D. Technical efficiency of the subsurface drainage on agricultural lands in the Moldova River Meadow. In: Kulshreshtha S, Elshorbagy A, editors. Current Perspective on Irrigation and Drainage. London: IntechOpen. DOI: 10.5772/67258

[23] Alcamo J. Water quality and its interlinkages with the sustainable development goals. Current Opinion in Environmental Sustainability. 2019;**36**:126-140. DOI: 10.1016/j.cosust.2018.11.005

[24] European Commission: Directive on environmental quality standards in the field of water policy (Directive 2008/105/EC) [Internet]. 2008. Available from: https://eur-lex.europa.eu/legal-content/EN/TXT/?uri=CELEX%3A32008L0105 [Accessed: July 20, 2021]

[25] United States Environmental Protection Agency: Water Topics [Internet]. 2019. Available from: https://www.epa.gov/environmental-topics/water-topics [Accessed: July 20, 2021]

[26] Copernicus-Corine Land Cover [Internet]. Available from: https://land.copernicus.eu/pan-european/corine-land-cover/clc2018 [Accessed: July 20, 2021]

[27] Dunea D, Iordache S, Liu HY. et al. Quantifying the impact of $PM_{2.5}$ and associated heavy metals on respiratory health of children near metallurgical facilities. Environmental Science and Pollution Research. 2016;**23**:15395-15406. DOI: 10.1007/s11356-016-6734-x

[28] Iordache S, Dunea D, Ianache C. Optimizing nutrients removal from municipal wastewater treatment plants using computer simulation. Revista de Chimie. 2015;**65**(2):242-248

[29] Negm AM, Romanescu G, Zeleňáková M. Water Resources Management in Romania. Berlin,

Heidelberg: Springer International Publishing; 2020

[30] Giri S. Water quality prospective in twenty first century: Status of water quality in major river basins, contemporary strategies and impediments: A review. Environmental Pollution. 2021;**271**:116332

[31] National Administration of Romanian Waters. The Management Plan of the Argeș-Vedea Hydrographic Area [Internet]. Available from: https://rowater.ro/wp-content/uploads/2021/03/PMB-ABA-Arges-Vedea_actualizat2016-2021-1.pdf [Accessed: July 20, 2021]

Chapter 2

A Critical Analysis of the Water Quality Impacts on Water Resources in the Athi River Drainage Basin, Kenya

Shadrack M. Kithiia

Abstract

This paper presents a critical analysis of the water quality changes and their impacts on water resources within the Athi River Catchment and its implications of the people's livelihoods. The paper analyses the effects of land use activities on water quality in the headwater areas of the basin which has a profound impacts on the downstream water uses within the basin. The paper in addition makes an attempt to relate the impacts of human activities on water quality degradation trends within the basin against the available and potential water resources in the basin. The Athi River is the second largest in Kenya and traverses areas of diverse land use activities from the more agricultural head water areas through the industrial hub of Kenya in Nairobi to the Indian Ocean discharging its waters near Malindi town north of Mombasa. The paper gives a detailed analysis of the impacts of human activities on the water resources in regard to water quality degradation, pollution and mitigation measures. The study was based on field data collection and measurements and laboratory analysis. The researcher used 10 sampling points located within the Nairobi sub-basin and distributed along the river profile to examine the trends in water quality degradation and its implication on human livelihoods in the basin. The researcher noted a declining trend in water quality status downstream the river profile. A close analysis of the water situation in the basin paints a blink future on the available water resources in the basin against the projected water uses and increasing population compounded by the impacts of water pollution and climate change. This is likely to increase incidences of water shortage and food insecurity in many parts of the basin. The researcher recommends more investments in water harvesting infrastructure, environmental conservation and adoption of modern water management technologies.

Keywords: Water resources, water quality critical analysis, livelihoods, increasing water demands

1. Introduction

The Athi River drainage basin is the second largest basin in Kenya after the Tana River drainage basin [1]. Furthermore, the river traverses agro-ecological zones of diverse climatic characteristics and land use activities. The river draws most of its headwaters from the Kenya Highlands (Kikuyu and Ondiri springs) flowing through the dry and semi-arid lands of Kenya and discharging its waters into the Indian Ocean, north of Malindi town.

In addition, the Kenya highlands are endowed with vast water resources but land use activity changes in these areas have led to water quality degradation in addition to affecting the flow characteristics of the rivers in the basin. Within the basin, there are two major urban and industrial centres from Kenya. Nairobi city, which lies at the upper catchment areas, and Mombasa, which is located on the southern outlet of the basin. Most of the upper-Athi river tributaries and the three main streams investigated (Ngong, Nairobi and Mathare) drain the Kikuyu escarpment through the city (now Nairobi County) and joining before the Kilimambogo hill (Donyo Sabuk) to form the main Athi-Sabaki River as shown in **Figure 1**.

The Athi River Catchment area borders the Tana Catchment area in the north and is located in the southern part of the country as shown in **Figure 1** and borders the Indian Ocean in the east, Republic of Tanzania in the south, and the Rift Valley in the west. In addition, the Aberdare Range, one of the Five Water Towers, lies in the northern edge of the area. The basin occupies a total area of 58,639 km^2, which corresponds to 10.2% of the country's total land area. Based on the Kenya National Bureau of Statistics report [2] the recent Census 2009, population of the basin in 2010 was estimated at 9.79 million people or 25.4% of the total population of Kenya with a population density of 167 persons/km^2.

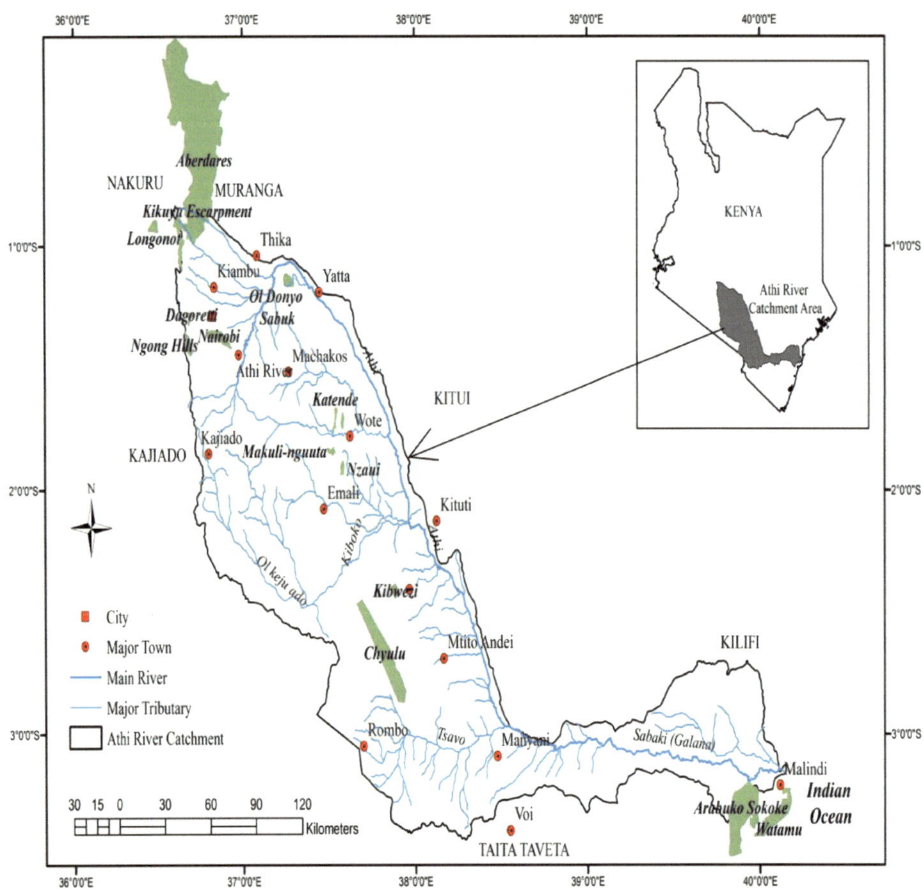

Figure 1.
Athi River catchment basin and main tributaries.

The catchment basin varies in topography ranging from 2,600 masl in the Aberdare Range to the sea level in coastal area. The basin covers three zones in altitude, with the upper zone at 2600–1500 masl, middle zone at 1,500–500 masl, and coastal zone at 500–0 masl.

The upper catchment area (headwaters area) is the centre of commercial production with many industrial and agricultural activities, which discharges their waste waters onto the river system. This pollutes the water that is used further downstream for irrigation purposes and rural domestic water supply. The continued discharges of effluents into the river system are a major concern for sustainable agricultural farming in the downstream areas and hence affecting the peoples' livelihoods and food security.

2. Study area characteristics

The Athi River flows from the southeast and north-eastward in the upstream reaches of the City County of Nairobi and then turning to the southeast in the north of Ol Doinyo Sabuk hill and flows along its boundary with Tana River drainage basin. The total drainage area is about 37,750 km^2 representing 64.4% of the Athi River Catchment Area. Some of its tributaries such as Lumi River, Lake Jipe, and Lake Chala flow into Tanzania and the Umba River flows from Tanzania to Kenya. Others such as the Rare, Mwachi, Pemba, and Ramisi rivers flow into the Indian Ocean making a total drainage area of 19,493 km^2. Several springs exist in the basin such as Mzima, Kikuyu, Njoro Kubwa, Nol Turesh, etc., whose waters are used both for Municipal and domestic water supply.

The mean annual rainfall in the basin ranges between 600 mm in the central part of the area to 1,200 mm in the upstream area of the Athi River with an overall average mean annual rainfall of 810 mm. The JICA [1] report calculated the renewable water resource, which is defined as precipitation minus evapotranspiration, at 4.54 BCM/year in 2010 for the basin and per capita renewable water resources at 464 m^3/year/capita. This is an indication that the basin is tilting towards a water scarcity scenario since the per capita value is less than the global accepted value of 1000 m^3/year/capita (UN [3, 4]). According to the World Bank Survey [5], Kenya's renewable internal freshwater resources stood at 412cubic metres per capita.

The main land use activities in the basin include urban, residential, industrial, transport in the main urban centres (Nairobi and Mombasa) and agricultural and livestock keeping both in the middle and lower reaches of the river as shown in **Figure 2**. These various land use activities have significant impacts on the water resources in the basin. They cause significant water quality degradation in the basin, reduce the available water resources in the basin while increasing competition in water usage and conflicts. The overall implication is reduced water availability for various uses in the basin and hence affecting people's livelihoods potential and leading to increased vulnerability to climate change. This calls for better adaptive strategies to cope with reducing water availability and re-innovation of sustainable methods of water resources exploitation and use.

These various land use systems contribute significantly to pollutants, pollution and water quality degradation as well as to changes in river hydrology, which is worth investigation. The land use changes in a spatial manner from the rich agriculturally based system, through residential and urban to industrial, making it ideal for investigation of gradual water quality degradation downstream the river systems. The present study on land use changes and their effects on the water quality and livelihoods in the drainage basin is an eye opener to the problems related to the land-water nexus and development in the county and country in general [6].

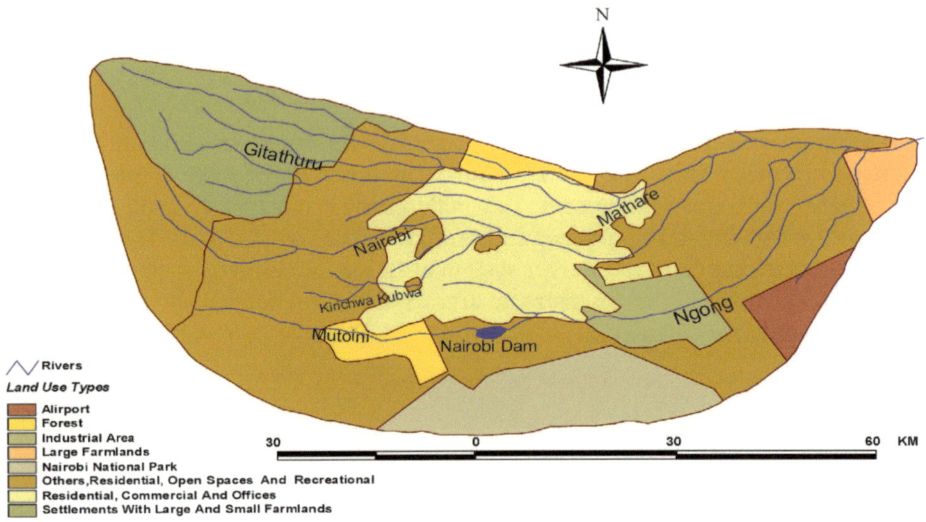

Figure 2.
Land use systems in the upper catchment area (Nairobi sub-basins). Source: Field data (2006–2017) and adapted from Kithiia [6].

3. Hydrology of the Athi River drainage basin

The Athi River basin is drainage basin number three (3) of the Kenya's drainage basins and is about 540 km long. The river drains a catchment area of about 70,000 km^2 (66,837 km^2) representing 12% of Kenya's total land area. The mean annual run-off is about 1294×10^3 with annual rainfall of 550 mm translating to 19 $m^3 s^{-1}$ mean annual runoff for the whole basin [6, 7].

The Athi River originates and drains the southern slopes of Aberdare ranges and comprises the southern part of the country east of the Rift valley. It covers large parts of Kiambu, Nairobi, Machakos and Makueni Counties. The river flows through a country of basement complex rocks, the areas being mostly semi-arid and subject to long drought periods. Its tributaries from the highlands flow in deep valleys close together, almost forming a parallel drainage system.

The main tributaries are Ruiru and Ndarugu. Other minor tributaries but of great hydrological importance are Ngong, Nairobi, Mathare, Mbagathi, Riara and Gitathuru. Ngong, Nairobi and Mathare tributaries drain the upstream areas and traverse the Nairobi city and its environs. After being joined by its tributaries, the Athi River flows down steeply in a series of falls and rapids in a metamorphic formation. It goes down reducing in capacity due to underground seepage as a result of geological configuration; e.g. 95 $m^3 s^{-1}$ above Kwaa, 65 $m^3 s^{-1}$ at Kibwezi (161 km) downstream, 40 $m^3 s^{-1}$ 70 miles (113 km) above the confluence of Tsavo (Republic of Kenya, [8]). Further down, it becomes reinforced by Tsavo River from Mt. Kilimanjaro and Mzima springs of the Chyulu hills which provide dry weather inflows after where it changes its name into Galana or Sabaki and flows in a series of meanders over sandy beds until it drains its waters to Indian Ocean north of Malindi as indicated in **Figure 1**.

The river carries with it an enormous volume of suspended sediments because of erosion and other human activities in its upstream reaches. The total amount of sediments discharged by the river into the Indian Ocean is estimated at 2,057,487 tonnes/year as indicated in **Table 1**. The heavy sediment loads discharged into the Ocean are responsible for the highly colored beaches of Malindi, which have changed totally to brownish therefore affecting the tourism industry by polluting

Code	River	River Profile section	Suspended Load	
			Mean (ppm)	Annual (t yr.$^{-1}$)
3AA04	Mbagathi	Upper drainage area (Headwaters)	193	4,456
3BAA22	Nairobi		57	2,231
3BB10	Riara		118	1,474
3CB05	Ndarugu		202	29,356
BDA02	Athi (Twake Conf.)	Middle reaches	153	131,089
3F02	Athi (Tsavo)		549	753,627
3HA12	Athi (L.falls-mouth)	Lower reaches	859	2,057,487
Source: adapted from NWMP [9] report Vol. 1, Table 2.31 & [6]				

Table 1.
Suspended load (t yr.$^{-1}$) and its volume of some selected streams in the Athi river drainage basin.

the sand beaches around Malindi town. The heavy sediment loads further adversely affect marine life by reducing light penetration and fishery. This in turn affects local fishing activities and livelihoods by reducing fish catches and sales, which the local communities relies on. Kithiia [7], Aketch and Olago [10] and Mavuti, [11], highlighted an increasing trend in water quality degradation within the basin. This complemented the findings of the present study.

4. Study basin characteristics

4.1 Water resources and water demands

A critical analysis of water resources in the basin indicates that water resources are affected by the different water use demands and their allocations. This indicates a close relationship between the available water resources against various water use activities, and how the water resources are allocated. This minimizes the possibility of water depletion and scarcity in the basin to enable a sustainable use of the water resources. Water resources in the Athi Catchment Area (ACA are mainly used for agricultural production (irrigation), domestic water supply, industrial uses and municipal uses in the several urban and municipal towns in the basin. The JICA [1] report summarized the available water resources as estimated in basin for year 2010 and projected for year 2030 as presented in **Table 2**.

The annual water demands estimated for the year 2010 and projected for 2030 in the basin are further summarized in **Table 3**. The projection given for 2030 followed the national development targets of Kenya Vision 2030 and socioeconomic framework [1], but without considering the available water resources.

The present study used ratios of available water resources and water demands and water deficits to show increasing water deficit trends due to increasing water demands in the basin. These are shown in **Tables 3** and **4** respectively. This will have far reaching implications on peoples' livelihoods as trends in decreasing water resources are likely to affect human activities such as irrigation activities and other water related production activities.

The present study notes that the water demands of 1,145 MCM/year in 2010 as presented in the JICA [1] report are equivalent to 76% of the available water resources (water stress ratio) as indicated in **Table 4**. This ratio far exceeds the severe water stress ratio of 40% indicating that the Athi catchment area is already

Year	Surface Water	Groundwater	Total
2010	1,198	305	1,503
2030	1,334	300	1,634
Ratio of 2030 to 2010	111%	98%	109%

Source: JICA Study Team [1].

Table 2.
Estimated and projected annual available water resources (ACA) in the Athi river basin (unit: MCM/year).

Year	Water Demands (MCM/year)						
	Domestic	Industrial	Irrigation	Livestock	Wildlife	Fisheries	Total
2010	519	93	498	25	3	7	1,145
2030	941	153	3,418	59	3	12	4,586

Source: JICA Study Team [1].

Table 3.
Estimated and projected water demands by sub-sector (ACA) in the study basin.

Description	2010	2030
Available Water Resources (MCM/year)	1,503	1,634
Water Demands (MCM/year)	1,145	4,586
% Water Demands/Available Water Resources	76%	281%
Water Deficits (MCM/year)	745	4,153
% Water Deficits/Water Demands	65%	91%

Source: JICA Study Team [1].

Table 4.
Ratios of water demands/available water resources and water deficits/water demands (ACA).

experiencing severe water stress. The expected water stress ratio is projected to increase to 281% by 2030. The estimated water deficits of 745 MCM/year in 2010 are projected increase to 4,153 MCM/year by the year 2030 as indicated in **Table 4**. The study envisages that the water demands that can be covered by the available water resources are as presented in **Table 5** taking into account the allocated amounts of the surface water and groundwater to satisfy the 2030 water demands as given in the JICA [1] baseline report.

The present study notes that due limited water resources and to afford a viable irrigation system and sustainable food production in the basin and enhanced livelihoods, the projected target of new irrigation development area of 233,628 ha for 2030 need to be reduced to 46,108 ha based on the water resources allocation plan and used as the data base for the future water resources management plan of the Athi catchment area [1]. In general, the Athi basin is likely to experience enormous water shortages within the projected period. This scenario compounded with the increasing trends of water demands, intensive irrigation projects, competition in water uses and increasing water quality degradation means that proper water use strategies should be put in place to avert a looming danger of severe water scarcity in the basin. This is likely to have negative impacts on the people's livelihood in the basin since majority relies on rain fed agriculture.

Subsector	Water Demand (2030)	Water Resources Allocation	
		Surface Water	Groundwater
Domestic	941	819	122
Industrial	153	77	76
Irrigation	917	*882	35
Livestock	59	59	0
Wildlife	3	3	0
Fisheries	12	12	0
Total	2,085	1,852	233

**Including water demand to be supplied by water resources of Tanzania of 154 MCM/year.*
Source: JICA Study Team [1].

Table 5.
Water resources allocation plan for water demands in 2030 after balance water study (ACA) (unit: MCM/year).

5. Study methodology

The present study employed both field data collection and secondary data sources to advance the researcher's opinion that due to the increased human activities and water quality degradation, the basin is likely going to experience severe water shortages and scarcity (hence affecting people's livelihoods).

Field and laboratory analysis and observations, standard empirical formulae for the determination of suspended sediment loads and discharge measurements were used to analyze the collected data. The used quantitative methods included principal component analysis, time series and chemical methods in determination of the water quality status.

Data on water quality and river discharge was necessary as well as the information on land use changes over time and space. This is relevant to the study since it is a factor associated with changes in river flow characteristics and quality degradation. Monthly data on river flow characteristics and water quality parameters were used, some of which were measured in the field during the period of investigation. Water sampling points indicated in **Figure 3** were used both for field samples and river gauging stations for secondary data.

The collected water samples were analyzed in the laboratory to quantify the sediment loads and the water quality deterioration changes along the river profiles. The total suspended sediments concentrations (TSSC) were determined on two replicate samples by gravimetric method according to Mcgrave [12] and Woodroffe [13]. The filtration was carried out by suction in which two vacuum flasks were run off one pump with open filter holders being mounted on each flask. After the determination of the water volume, sediment-water mixture was filtered through pre-weighed Whatman GF/C filters (4.7 cm wide, pore size 0.4 μm, thickness 260 μm) and kept in individually numbered aluminum packs. These were handled at the edges using flat-bladed tweezers. Following the filtration, the filtrates were washed with filtered distilled water and dried in oven at a temperature of 105°C for 24 hours. After removal from the oven, the filters were left to cool to room temperature for about 2 hours before they were reweighed using a sensitive electronic balance to the nearest 0.0001 g. Whatman GF/C filters of 47 mm diameter were used because they have a high flow rate and take a high sediment load [12]. Heavy metal analysis for mercury, chromium, cadmium, zinc, nickel, copper, iron,

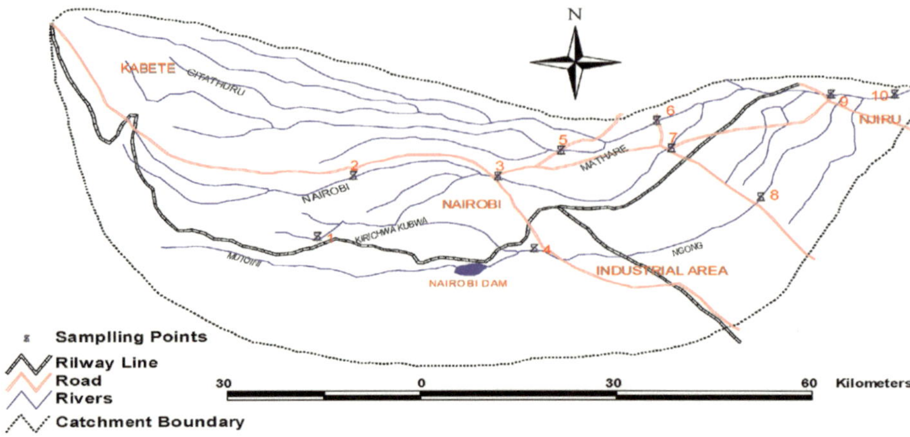

Figure 3.
Water sampling points in the Nairobi River sub-basin (upper Athi catchment basin).

manganese and fluoride was done to establish the concentration variations along the stream course distance-wise before and after the rivers pass through the city of Nairobi (urban/commercial areas) or part of its environs using the same method.

Water samples for water quality determination were collected in a depth integrated manner at the middle of the river. Each sample was taken according to the standard laboratory analysis and the concentration of each parameter determined. The basic method applied in the determination of each metallic cations was the Atomic absorption spectrometry (AAS). Measurement of other water quality parameters was done using standard laboratory methods and included BOD_5, COD, TSS, pH, Total Alkalinity, Total dissolved solids, Conductivity, Calcium, Magnesium, Sodium, Potassium, Chloride, Fluoride, Total hardness and Turbidity.

BOD_5 was determined over 5 days for biochemical oxidation of organic substances at 20°C. The detection limit was set at about 5 mg l^{-1} and the same was done for the chemical oxygen demand (COD). The total dissolved solids (TDS) concentration was done through evaporating the water sample on a previously weighed dish and the residue dried at 180°C and then weighed again to allow for the concentration to be determined while suspended solids (SS) were determined by filtration of a well mixed sample on a standard glass-fiber filter disk (0.45 mm filter; detection limit SS $\leq$5 mg l^{-1}). Fluoride was determined by titration of the water sample with a standard Thorium nitrate in a solution buffered at about 2.9–3.3, using sodium alizarin sulphonate as indicator. Volumetric analysis or otherwise filtration method was used in the analysis of most of the other water quality parameters since it is usually more rapid than the gravimetric analysis if the filtrate reagent is specific for the water quality constituent. The sensitivity and/or precision may also exceed that of a gravimetric analysis for some determinants. It was widely used in the present study for the analysis of water quality parameters as pH, total hardness, and alkalinity according to Mancy [14]. Their results were used to group the parameters into three main categories; chemical, biological and physical related water pollution parameters. In addition, their magnitude values (concentrations) were used to infer on whether there was water quality deterioration or not.

6. Sediment measurement

Water samples for suspended sediment analysis were collected at the middle and both sides of the river banks using USDH48 sampler. At each point, a depth integrated (equal transit rate) water-sediment mixture samples were taken. The samples were stored in a 3 liters sample bottle and taken to the laboratory for suspended sediment concentration determination according to American Public Health Association [15–17] techniques and methods. Sediment discharge (mg l^{-1}) at the cross-section was computed by multiplying suspended sediment concentration (mg l^{-1}) with the river discharge $m^3 s^{-1}$), and the correction factor of 0.0864 [18]. The total sediment yield in tonnes for the days (n) sampled was computed according to Jorgensen and Vollenweider, [19] and Sharma [18] as:

$$S = 0.0864 \sum_{i=1}^{n} C_i\, Q_r \tag{1}$$

Where:

S is the sediment yield in t day^{-1}, C_i is the measured suspended sediment concentration in mg l^{-1} and Q_r is the cross sectional river flow in $m^3 s^{-1}$. This procedure was used to calculate the sediment yields and mass loadings in the sub-basins investigated and as means for sediment loading downstream the river profile.

7. Results and discussions

This sections details the results of the field data observations and the laboratory analysis on the water sampled collected. It was intended to give an overall picture of the water quality status against the available water resources, human activities and land use changes in the basin.

The results in **Table 6** and **Figure 4** clearly indicate a close link between river flow characteristics and rainfall patterns within the basin and the entire catchment area. The high flows were noted to occur in the months of April–May and replicated in the months of October to December corresponding quite well to the rainfall patterns in the basin.

From **Table 7**, it can be observed that downstream the river profiles, the amount of total sediment load increases. The exception only occurs at Njiru 2 sampling point, which is the outlet of all the three streams investigated where the value of TSS was calculated to the value of 5166.23 t $year^{-1}$. This was attributed to widening of the river channel, sedimentation and dilution effects, which may have resulted to the deposition of the sediment loads and combined stream discharge increase from the three streams at this point.

Sub-basin River	Mean monthly river flows per sub-basin in $m^3 s^{-1}$ within the upper catchment area											
	Jan	Feb	Mar	Apr	May	Jun	Jul	Aug	Sep	Oct	Nov	Dec
Nairobi	1.3	0.6	1.6	2.8	2.7	3.6	1.7	0.8	0.7	1.9	5.8	2.9
Ngong	0.2	0.4	0.6	0.8	1.0	1.4	0.7	1.3	0.8	0.5	1.2	0.3
Mathare	1.0	0.2	0.9	1.0	1.1	0.8	—	—	2.1	0.9	2.3	1.8

Source: field data (1998–2017).

Table 6.
Measured mean monthly river discharge sub-basin^{-1}.

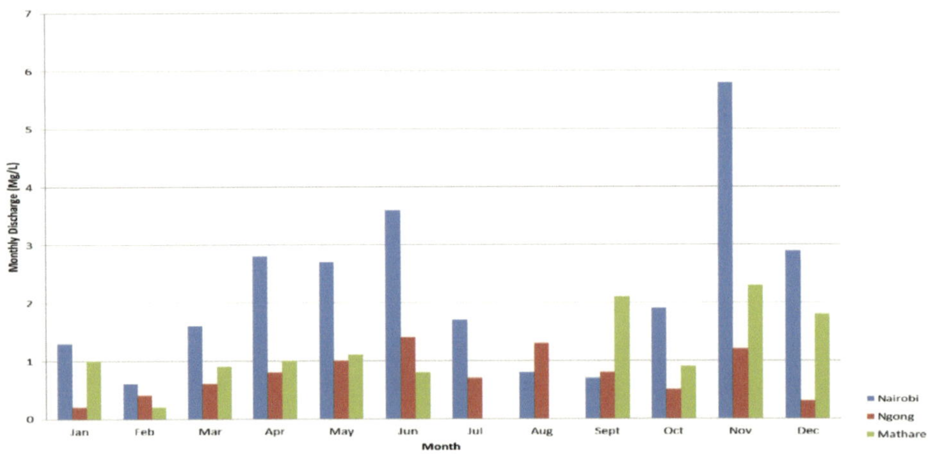

Figure 4.
Mean monthly river discharge per sub-basin in the upper sub-basins (m^3/s). Source: filed data (1998–2017).

Sampling Stations	Station Number	Main river	TSS Values (t $year^{-1}$)
Langata rd Bridge	4	Ngong	189
Embakasi rd Bridge	8		1853
Thika rd Bridge	5	Mathare	1295
Outering rd Bridge	6		3167
Muthangari	2	Nairobi	989
Museum	3		1855
Outering road	7		6618
Njiru 1 after Mathare Confluence	9		14720
Njiru 2 All streams joined	10		5271
Source: field data (1998–2017).			

Table 7.
Mass loadings of total suspended sediments (TSS) in the Nairobi river sub-basins in tonnes per year.

8. Impacts of land use activities on river hydrology

Land use changes in the upper catchment areas through the city to the lower areas were observed as the causes of water quality degradation [6, 20]. This is further attributed to increased storm water, reduced infiltration rates and hence flooding in the urban areas.

Poor land use and management impacts on the hydrology of the catchment resulted to surface run-off increases which then determined the increased discharge of the rivers. In the sub-basins investigated and downstream the Athi River Catchment area, physical water quality parameters increased with discharge downstream the river courses while metallic ions decreased or increased depending on river sub-basin as shown in **Table 8**.

The study noted a general trend of increase in total suspended sediments with increase in river volume (discharge) and consequently more pronounced water turbidity. The reverse, it may be attributed to river widening in channel size and more settling or deposition of the sediments. In addition, the study noted that away

Sampling point	River	Mean Concentrations				
		Q ($m^3 s^{-1}$)	TSS ($mg l^{-1}$)	COND (μcm^{-1})	TDS $mg l^{-1}$)	TUR (N.T.U.)
Muthangari	Nairobi	0.772	157.6	392.1	239.7	69.4
Museum	Nairobi	1.376	129.4	397.7	244.2	69.3
Outering Rd.	Nairobi	2.140	255.7	564.5	290.9	65.5
Njiru 1	Nairobi	5.083	199.2	509.5	310.9	67.8
Njiru 2 (10)	Nairobi	5.341	95.5	474.8	298.8	28.5
Thika Rd.	Mathare	0.738	161	352.2	215.5	35
Outering Rd.	Mathare	1.371	251.2	527.1	349.9	85
Kibera slums	Ngong	0.110	164	233	88	98
Langata Rd.	Ngong	0.305	59	598.8	59	42
Embakasi	Ngong	0.949	180	611.7	174.4	71

Source: field data (1998–2017); COND-Electric conductivity, TDS-Total dissolved solids, TSS-Total suspended sediments, TUR-Turbidity.

Table 8.
Mean measured values of physical water quality parameters at various sampling points.

from the main land use activities, water in the streams was found to be less polluted both physically and chemically. This implies much of water quality degradation in the upper catchment areas was attributed to changes in land use activities.

9. Present water supply situation of in the Athi River catchment area (ACA)

Data from the KNBS [2] indicates that the population of ACA in 2010 stood at 9.79 million including an urban population of 6.51 million and a rural population of 3.28 million. This population is concentrated in both Nairobi area and Mombasa area, which are the major metropolitan areas in the basin. A study by the JICA team [1] and adopted by this present study estimated the current situation of water connection and supply of ACA as shown **Table 9**.

Table 9 clearly indicates that 24% of the population in the basin gets water from unregistered water vendors, streams, lakes and ponds without proper treatment, which are designated as an unimproved drinking water sources. Around 22% of the population gets water from springs, wells or boreholes, which are considered safe for water supply. A considerable proportion of the population (54%) is supplied with water by water service providers through pipes indicating a good trend in

Type	Piped by WSPs	Spring/Well/ Borehole	Water Vendor	Stream/Lake/Pond/ Others
Urban Population	63%	17%	17%	3%
Rural Population	28%	34%	3%	35%
Total Population	54%	22%	13%	11%

Source: JICA Study Team based on Census 2009 data [1].

Table 9.
Current situation of water connection (ACA).

water provision in the basin. However, with a projected urban population of 11.22 million and decrease of rural population by 0.47 million as a result of rural–urban migration in search of better living conditions and employment, water provision and supply remain a big challenge.

The urban population in the basin is well served with piped water representing 63% and this ratio is the highest in all six catchment areas. The implication of this is that the catchment requires implementing a large-scale urban water supply system development to cope with the urban population increase of 1.08 million and achieve the target coverage ratio of 100% as envisaged in the JICA study team report. It also calls for a proactive water resources development strategy that includes investments in water harvesting and conservation of water catchment areas through re-afforestation programmes.

10. Water resources development strategy

The ACA is divided into three areas in terms of water supply zones, such as Nairobi surrounding area, Mombasa surrounding area and other area for urban water supply systems (UWSS) considering the characteristics of the three areas. The present study, which is in line with the JICA report, noted that in overall Urban Water Supply Systems (UWSS) are planned for 32 Urban Centres (UCs) within the basin. Out of the 32 UCs, 16 are planned for Nairobi and satellite towns and nine UCs in Mombasa and coastal surrounding area while one water supply system is planned to cover several UCs.

The projected water supply capacity required for UWSS in ACA in 2030 is 2,260,000 m^3/day against the current water supply capacity (including capacity under construction) of 699,000 m^3/day. This implies that, additional capacity of 1,560,000 m^3/day need to be developed by 2030. This is proposed to be undertaken through the following three types of projects as recommended in the JICA [1] report.

a. Rehabilitation of existing UWSS: This is to be achieved by installing water meters in all households and replacement of old pipes of existing UWSS of the 30 UCs. In addition, the rehabilitation shall include repair and replacement of mechanical and electrical equipment in water treatment plants and pumping stations.

b. Expansion of UWSS: The projected total planned capacity of expansion is 1,542,000 m^3/day in 28 UCs out of the above 29 UCs to meet the water demand in 2030.

c. Construction of new UWSS: The total capacity of new construction is estimated at 19,000 m^3/day.

d. There are 31 plans of urban water supply development projects to cover 21 UCs and surrounding areas in ACA with an estimated 1,215,000 m^3/day of total water supply capacity. This is planned to augment urban water supply systems in the basin.

e. Proposed Water Supply Development Plan.

The development strategy of water resources in ACA for the proposed UWSS is presented in **Table 10**, while those proposed for LSRWSS and SSRWSS are in

Type of Project		Target Area	Total Capacity (m^3/day)	Service Population (million persons)
Urban Water Supply	Rehabilitation	30 UCs	699,000	17.01
	Expansion	29 UCs	1,542,000	
	New Construction	2 UCs	19,000	
	Total	32 UC	2,260,000	
Rural Water Supply	LSRWSS	10 Counties	209,000	4.04
	SSRWSS	10 Counties	110,000	
	Total	10 Counties	319,000	

Note: The water supply development plan of ACA includes Thika with 0.51 million population in 2030. Thika is located in TCA, but it has been covered by water supply system in ACA. LSRWSS- Large Scale Rural Water Supply System and SSRWSS- Small Scale Rural Water Supply System.
Source: JICA Study Team [1].

Table 10.
Proposed water supply development plan (ACA).

Items		Urban Water Supply	Large-scale Rural Water Supply	Small-scale Rural Water Supply	Total
Service Population (million)	2010	5.29		2.15	7.44
	2030	17.01	2.04	2.00	21.05
Water Supply Capacity (m^3/day)	2010	699,000	100,000	108,000	907,000
	2030	2,260,000	209,000	110,000	2,579,000
Operating Body		Registered WSPs	Registered WSPs	Individual, Community, etc.	—
Target Towns/ Areas		32 UCs	10 Counties		—

Source: JICA Study Team [1].

Table 11.
Water supply situation in 2030 (ACA).

Table 11. This further demonstrates the situation of urban centers subject to urban water supply system development.

The proposed water supply development plan for ACA is outlined **Table 10**.

It is anticipated that by using the water supply development plan indicated in **Table 10**, the water supply situation of ACA in 2030 will be improved as shown in **Table 11**.

The present study proposes the construction of the eight and five dams within the ACA and TCA respectively as noted in the JICA report [1]. This coupled with the expansion of the inter-basin transfer system from the Tana River Catchment Area (TCA) will ensure adequate water supply to both Nairobi city and its environs will be met.

Three new dams in ACA and expansion of two existing intra-basin water transfer systems are envisaged to increase water supply systems in Mombasa city and the coastal area. More dams are further proposed to cover other areas within the Athi River catchment area (ACA). These efforts are likely to increase land under irrigation within the basin and further help in the improvement people's livelihoods in the basin.

11. Current situation of water resources development

The Athi River basin with a total area of 58,639 km^2 receives an annual rainfall of 810 mm, which is categorized between rather high rainfall of around 1,300–1,400 mm in LVNCA and LVSCA and low rainfall of around 500 mm in RVCA and ENNCA. Rainfall amounts differs spatially within the basin ranging from 500 mm in the southern part near the border with Tanzania to about 1,200 mm in the western part of the country.

As of 2010, the estimated water resources in the basin were to the tune of 1,198 MCM/year for surface water and 333 MCM/year for groundwater while the water resources estimated for 2030 stands at 1,334 MCM/year considering the effect of climate change, while the available groundwater resources is 303 MCM/year which is almost same as the amount of 2010.

On the other hand, water demands in the basin is estimated to be 1,145 MCM/year, which consist mainly of domestic water demands for the population of 9.79 million and irrigation water demands for the irrigation area of 44,898 ha. Most of the population is concentrated in and around Nairobi and Mombasa cities and its area along the coastal belt. This implies that there are increased water demands in these urban centres. The study noted that the existing water resource structures/facilities would not be able to satisfy the greatly increased 2030 water demand because of the uneven distribution of water resources both spatially and temporally. Therefore, new water resources structures/facilities are required to be developed.

More attention needs to be paid to the domestic water supply in Nairobi and satellite towns, and also in Mombasa and coastal areas where future domestic water demands will increase drastically, but the available water resources are limited.

The Strategies for the water resources development in Athi River basin should formulate a well-balanced development plan between water resources and demands, based on the current situation of the catchment area and future water demands. Theses should include:

a. Development and expansion of inter-basin water transfer facilities

b. Construction of both large and small scale dams to augment the existing water sources in the basin

c. Exploration and drilling of boreholes in the basin

d. Investing in rural water harvesting facilities to reduce the pressure on the existing water supply systems and to ensure sustainable water supply in the basin.

12. Water resources use implications in the Athi River basin

This paper has attempted to broadly analyze the available water resources within the Athi River catchment Area (ACA) highlighting the problems associated with water resources use and demands against availability, potential water uses and proposed projects in the basin, strategies to address the scarcity of water in the basin, water pollution trends and their implications on water resources in the basin. An attempt to assess the hydrology of the basin was also performed to give an overview of the present and the projected water use demands and use trends in the catchment area. Evidence of the declining trends in water availability against the various uses was demonstrated in the paper. The results presented in this paper emphasize that improved water supply systems and the expansion of the existing water sources will reduce the pressure on the existing water resources from various water users in the basin. Overall, this is likely to improve and augment the available water and reduce conflicts and further improve on people's livelihoods.

13. Conclusion

The paper concludes that the Athi River Catchment Area (ACA) is endowed with enormous water resources but which is mal-distributed in both space and time. The basin is home to the two major urban centres in Kenya; Nairobi and Mombasa which have seen an increasing demand on water resources as a result of increasing rural–urban migration of the population in search of employment opportunities. The water resources in the basin are faced with increasing water pollution and quality degradation from the various human activities, which affects the various uses of the water. The water resources are being used for irrigation purposes in the lower reaches of the river and industrial and commercial purposes in the upper reaches. These activities generate pollutants that pollute the water and hence reducing its usefulness and in turn affecting the people's livelihoods. The paper notes that efforts by both the national and county government are likely to increase the water supply by investing in water storage structures to augment the water resources in the basin. More irrigation land is also earmarked for expansion implying more opportunities for the residents and enhancing their livelihoods through growing of more value chain crops under irrigation systems.

The paper recommends more investments on water resources structures for rain water harvesting, reducing and monitoring of waste water discharges onto the present water sources. The county governments within the basin should be encouraged to invest in small scale water supply systems such as earth dams and shallow wells. More emphasis of environmental education on the conservation of water resources through new technologies for irrigation projects as well as manufacturing industries to avoid discharging of industrial effluents into the river system. Continuous water quality should be encouraged to ensure the water resources in the basin are not polluted and when harvested are not health hazards. Better land use systems in the headwater areas should be encouraged to avoid generation of sediments by erosion process. This in the overall will help in the sustainability of the water resources in the basin and enhance increased opportunities for the residents and improved livelihoods.

Author details

Shadrack M. Kithiia
Postgraduate Programme in Hydrology, Department of Geography and Environmental Studies, University of Nairobi, Nairobi, Kenya

*Address all correspondence to: skithiia@yahoo.com

References

[1] JICA. The development of National Water Master Plan, 2030. Draft Report: Kenya; 2012

[2] GOK-KNBS (2010). The National Population Census statistical data, Government Printer, Nairobi

[3] Ritchie H, Roser M. "Water Use and Stress". *Published online at OurWorld InData.org.* Retrieved from: 'https://ourworldindata.org/water-use-stress'. In: Online Resource. 2017

[4] UN Water. (2006). *Kenya National Water Development Report: A WWAP case study prepared for the 2nd UN world water development report: Water, a shared responsibility (2006). Un-Water/Wwap/2006/12.* Nairobi.

[5] World Bank. World Bank Survey on renewable internal freshwater resources, total (billion cubic meters). Aquastat data: Food and Agricultural Organization; 2017

[6] Kithiia SM. The effects of land use types on the hydrology and water quality of the upper-athi river basin, kenya, *unpublished Ph.D.* In: *Thesis* University of Nairobi. 2006

[7] Kithiia SM. Effects of Industries and Other land-use systems on the water quality within the Nairobi river sub-catchments. *Thesis* University of Nairobi: Kenya. *unpublished M.Sc*; 1992

[8] MOWD (MLRRWD). The Athi River Basin Integrated water resources management programme: Phase 1, Management of aquifers underlying the greater Nairobi area and monitoring of industrial pollution problems in the upper-Athi river basin. Draft final report. 1997

[9] MOWD. The National Water Master Plan. Ministry of Water Development, Kenya: Final report; 1992

[10] Aketch and Olago. Pollution Assessment in Nairobi River Basin. Pollution Assessment Report of the Nairobi River Basin. In: Africa Water Network. Okoth: P.F and Otieno P. (Eds); 2000

[11] Mavuti KM. Nairobi River Basin Project-Phase II, Pollution Assessment and Monitoring Sub-component, UNEP/UON. Kenya: Final Report, Nairobi; 2003. p. 2003

[12] Mcgrave IN. Suspended Sediment. In: *Dyer, R.K.(1979): Estuarine hydrography and sedimentation-* A handbook. Cambridge: Cambridge University Press; 1979. pp. 132-185

[13] Woodroffe CD. Studies of a Mangove basin, Tuff crater, New Zealand. In: *Biomass of Mangroves and Production of detritus*. Estuarine, Coastal and Shelf Science. 1985;**20**: 260-285

[14] Mancy KH. Instrumental Analysis for water pollution control. Ann. In: Arbor. Sciences Publisher: Inc; 1971

[15] APHA (1992) Standard methods for the examination of water and wastewater-18th Edition. A. Greenber, L.S. Clescer, A.D. Eaton (Ed). American Public Health Association, American Water Works Association, Water Environment, New York.

[16] APHA (1995) Standard methods for the examination of water and wastewater-18th Edition. A. Greenber, L.S. Clescer, A.D. Eaton (Ed). American Public Health Association, American Water Works Association, Water Environment, New York.

[17] APHA (2001) American Public Health Association, American Water Works Association, and Water Environment Federation, *Standard Methods for Examination of Water and*

Wastewater, 20th Ed., American Public Health Association, New York, NY.

[18] Sharma TC. *Some concepts about sediment rating equations*. In: Gichuki FN, Mungai DN, Gachene CKK, Thomas DB, editors. Land and Water management in Kenya: Towards sustainable land use. *The proceedings of the fourth (4th) land and water management workshop*, 15th-19th February. Vol. 1993. Kenya: Kikuyu; 1993. pp. 65-70

[19] Jorgensen SE, Vollenweider RA, editors. Guidelines of Lake Management Vol. Principles of Lake Management pg. 1988;**1**:53-69

[20] Kithiia, S.M. (1997) Land use changes and their effects on sediment transport and soil erosion within the Athi river drainage basin, Kenya. In: *Human Impact on Erosion and Sedimentation (proceedings of Rabat Symposium S6*, April 1997). IAHS Publ. No. 245, 1997. 145-150.

Chapter 3

Seasonal Variability of Groundwater Quality in Kapas Island, Terengganu, Malaysia

Mohmadisa Hashim, Arijatul Wardah Ahmad, Zahid Mat Said, Nasir Nayan, Hanifah Mahat, Yazid Saleh and Koh Liew See

Abstract

The chapter aims to evaluate the groundwater quality levels in Kapas Island, Terengganu, Malaysia during the monsoon changes of the Southwest Monsoon (SWM), Monsoon Transition (MT) and Northeast Monsoon (NEM) in 2018. Four locations were used for groundwater sampling namely, the Kapas Coral Beach Resort, Kapas Beach Chalet, Pak Ya Seaview Chalet, and Kapas Island Resort. Three water samplings at each station for every month in the monsoon. Six parameters of the Malaysian Water Quality Index (WQI), i.e., dissolved oxygen (DO), pH, biochemical oxygen demand (BOD), chemical oxygen demand (COD), total suspended solids (TSS) and ammoniacal nitrogen (NH_3-N), were used to evaluate the water quality. The findings showed the groundwater quality parameters are in Class I and II. However, according to WQI Malaysia, the water quality status during the three monsoons is slightly polluted. During the SWM, the WQI value was 76 (Class III), the MT was 77 (Class II), and the NEM was WQI 71 (Class III). Given this status, it requires more intensive water treatment as it is not suitable for direct drinking water supply. The implications of the study show that the quality of groundwater in Kapas Island has to improve by the tour operators.

Keywords: water quality index, groundwater, slightly polluted, water treatment, island tourism

1. Introduction

The increase in the population of an area will have an impact on the demand for clean water supply. When the demand for water supply exceeds the capacity, it will lead to a water crisis because the need for water is not only for domestic use but also for various human activities such as industry and agriculture. The main source of water to meet human demand comes from surface water sources such as rivers, ponds and lakes. However, due to increasing demand for water resources, the groundwater resources have been explored on a large scale, especially for drinking water supply [1–4].

The current trend of using groundwater resources as a domestic water supply for humans is no longer uncommon. Polluted river water resources have led the residents or responsible authorities in providing water supply to opt for groundwater resources [5, 6]. Obtaining groundwater resources may not be an obstacle in the plain or continental areas but this situation is slightly exacerbated in island areas. However, groundwater resources are susceptible to water pollution that resulted from human activities such as domestic waste disposal and industrial activities [7]. The main cause of groundwater pollution is due to the disposal of sewage waste from tourism activities, agriculture and residential settlement in an island area [5, 8–10] and at the same time also resulted from natural factors such as the encroachment of saltwater on groundwater or wells nearby the area [11–13].

Groundwater is generally a source of water that is clean and can be easily accessed by building a well and by using pipes and channeling it into the reservoirs. In this case, the lack of freshwater resources particularly in the island area has urged the residents living in the area to have to use this water resource. Therefore, these water resources are specifically for domestic use such as drinking water supply, cooking, bathing and washing. This situation is increasingly affected when the island has been turned into a tourist destination causing the demand for water supply to increase. Therefore, the status of this groundwater quality should also be given special attention so that it is safe to be used by tourists and the local community [7, 8, 14, 15].

Kapas Island in Marang, Terengganu is a less populated area due to its small land area, which is the case with other islands. However, this island has its own charm due to the beauty of its beaches, unpolluted sea and a fascinating marine park. However, the lack of surface water resources such as river water has led local people and operators of resort and chalet for tourism activities to use groundwater resources as the main source of domestic water supply. The major issue is that the condition of these groundwater resources are sometimes unclean, murky and suspended solids are found when channeled to tourist accommodations. Therefore, the purpose of this study is to analyze the quality of groundwater supplied by accommodation operators in Kapas Island to tourists and local residents by using the Water Quality Index (WQI) as set by the Department of Environment (DOE) Malaysia.

2. Study area and methods

2.1 Study area

Kapas Island is located in the district of Marang, Terengganu at latitude 05°13.042′N and longitude 103°15.700′E. Kapas Island was selected as a study area because of its active tourism activities and freshwater resources are highly needed for tourists' domestic use. The Marang district is famous for its Kapas Island and is one of the eight districts in the state of Terengganu. The entire area of Marang district covers a 666.54 km^2 area consisting of six sub-districts, namely Merchang, Pulau Kerengga, Jerung, Rusila, Bukit Payong and Alor Limbat.

Kapas Island is a short distance island located about 6 km away from Marang jetty and is one of the islands in the state of Terengganu (**Figure 1**). The size of Kapas Island is about 1.5 km^2 to 2.5 km^2 and is famous for its clear sea water, white sandy beaches and waving coconut trees, swings along the coast. The island is also quite secluded, away from the hustle and bustle of the mainland and known for its unique marine park in the island area filled with various soft and hard corals, fish, turtles and other interesting marine life making it a mandatory destination for scuba diving and snorkeling activities [16]. The relaxing atmosphere of this island gives comfort to visitors as well as the chances to try

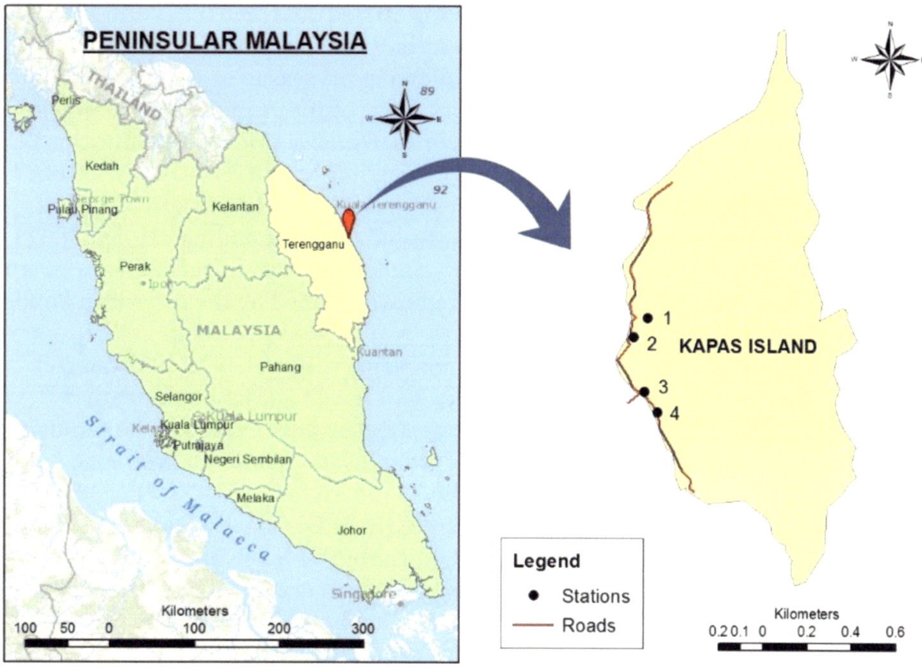

Figure 1.
Groundwater quality sampling station on Kapas Island.

Station	Station name	Latitude	Longitude
1	Kapas Coral Beach Resort	05°13.224′N	103°15.717′E
2	Kapas Beach Chalet	05°13.162′N	103°15.686′E
3	Pak Ya Seaview Chalet	05°13.018′N	103°15.707′E
4	Kapas Island Resort	05°13.004′N	103°15.742′E

Table 1.
The names and positions of groundwater quality observation stations.

out various water activities such as swimming, kayaking, waterboarding, hiking tropical forest tracks and climbing Bukit Singa (Singa Hill). In order to assess the level of groundwater quality in Kapas Island, four sampling locations were determined as shown in **Figure 1** and **Table 1**.

2.2 Study method

Water quality in Malaysia is measured by using the WQI set by the DOE Malaysia. According to Nurfadzlina et al. [17], this index is a measurement to give a comprehensive picture of the status of water quality for an area or water body. Meanwhile, Muhammad Fuad et al. [18] mentions that groundwater quality assessment consists of physical, biological and chemical parameters. In determining the status of groundwater quality in Kapas Island, only six water quality parameters were used as set by the DOE Malaysia in determining the quality status of a water body. The six parameters were pH, dissolved oxygen (DO), biochemical oxygen demand (BOD), chemical oxygen demand (COD), ammoniacal nitrogen (NH_3-N) and suspended solids (SS). In addition, two other parameters were also measured and analyzed, i.e., salinity (SAL) and the total dissolved solids (TDS).

Groundwater sampling in Kapas Island was carried out three times (August, October and November 2018). The three months were chosen to monitor for any possible difference in water quality levels following the season of monsoon winds, especially in Peninsular Malaysia. The observations of the groundwater quality in August represent the Southwest Monsoon winds, October (Monsoon Transition) and November (Northeast Monsoon winds). Sampling activities were carried out at four different locations and groundwater samples were tested in situ using YSI Multiparameter equipment. The parameters tested in situ including pH, SAL, TDS and DO. As for the other parameters, water samples were taken and analyzed ex situ (laboratory analysis), according to the procedure described by the American Public Health Association [19].

A quantitative approach was applied in analyzing the level of groundwater quality in Kapas Island. Descriptive statistics were used to describe the status of groundwater quality whether the observed parameters have exceeded the standards set by the DOE Malaysia. In this regard, the groundwater quality data obtained will be compared with the National Water Quality Standard for Malaysia (NWQSM) (**Table 2**), Water Uses and Classes (**Table 3**), DOE's WQI Classification (**Table 4**) and DOE's WQI Classification based on WQI (**Table 5**). Meanwhile, WQI formulas and calculations are shown in **Table 6**. The groundwater quality data of the four observation stations are presented in the form of tables and diagrams to provide an overall picture of the status of groundwater quality in Kapas Island.

Parameter	Unit	Class					
		I	IIA	IIB	III	IV	V
NH_3-N	mg/l	0.1	0.3	0.3	0.9	2.7	>2.7
BOD	mg/l	1	3	3	6	12	>12
COD	mg/l	10	25	25	50	100	>100
DO	mg/l	7	5–7	5–7	3–5	<3	<1
pH	—	6.5–8.5	6–9	6–9	5–9	5–9	—
Color	TCU	15	150	150	—	—	—
Conductivity	S/cm	1000	1000	—	—	6000	—
Floatables	—	N	N	N	—	—	—
Odor	—	N	N	N	—	—	—
Salinity	%	0.5	1	—	—	2	—
Taste	—	N	N	N	—	—	—
TDS	mg/l	500	1000	—	—	4000	—
TSS	mg/l	25	50	50	150	300	300
Temperature	°C	—	Normal +2°C	—	Normal +2°C	—	—
Turbidity	NTU	5	50	50	—	—	—
Fecal Coliform	count/100 ml	10	100	400	5000 (20000)a	5000 (20000)a	—
Total Coliform	count/100 ml	100	5000	5000	50000	50000	>50000

[a]*Maximum not to be exceeded.*

Table 2.
National water quality standards for Malaysia [20].

Class	Uses
Class I	Conservation of natural environment. Water Supply I - Practically no treatment necessary. Fishery I - Very sensitive aquatic species.
Class IIA	Water Supply II - Conventional treatment. Fishery II - Sensitive aquatic species.
Class IIB	Recreational use body contact.
Class III	Water Supply III - Extensive treatment required. Fishery III – Common of economic value and tolerant species; livestock drinking.
Class IV	Irrigation.
Class V	None of the above.

Table 3.
Classification of water quality and uses [20].

Parameter	Unit	Class				
		I	II	III	IV	V
DO	mg/l	> 7	6–7	3–5	1–3	< 1
pH		> 7	6–7	5–6	< 5	> 5
BOD	mg/l	< 1	1–3	3–6	6–12	> 12
COD	mg/l	< 10	10–25	25–50	50–100	> 100
TSS	mg/l	< 25	25–50	50–150	150–300	> 300
NH_3-N	mg/l	< 0.1	0.1–0.3	0.3–0.9	0.9–2.7	> 2.7
WQI		> 92.7	76.5–92.7	51.9–76.5	31.0–51.9	< 31.0

Table 4.
DOE's WQI classification Malaysia [20].

Sub index & water quality index	Index range		
	Clean	Slightly polluted	Polluted
Biochemical Oxygen Demand (BOD)	91–100	80–90	0–79
Ammoniacal Nitrogen (NH_3-N)	92–100	71–91	0–70
Suspended Solids (SS)	76–100	70–75	0–69
Water Quality Index (WQI)	81–100	60–80	0–59

Table 5.
Range of water quality index based on WQI [20].

3. Results and discussion

3.1 Groundwater quality based on parameters

Groundwater quality analysis in Kapas Island was done based on eight parameters, namely SAL, TDS, DO, BOD, COD, pH, NH_3-N and TSS. The findings of this study described the value of groundwater quality parameters based on the four designated study areas in addition to comparing the values obtained during the Southwest Monsoon (SWM), monsoon transition (MT) and Northeast Monsoon

WQI Formula	
WQI = (0.22*SIDO) + (0.19*SIBOD) + (0.16*SICOD) + (0.15*SIAN) + (0.16*SISS) + (0.12*SipH)	
where:	
SIDO = subindex DO (% saturation)	
SIBOD = subindex BOD	
SICOD = subindex COD	
SIAN = subindex NH_3-N	
SISS = subindex SS	
SIpH = subindex pH	
$0 \leq$ WQI ≤ 100	
Subindex DO (mg/l)	
SIDO = 0	$x \leq 8$
SIDO = 100	$x \geq 92$
SIDO = −0.395 + 0.030x2–0.00020x3	$8 < x < 92$
Subindex BOD (mg/l)	
SIDOD = 100.4–4.23x	$x \leq 5$
SIDOD = 108* exp.(−0.055x) – 0.1x	$x > 5$
Subindex COD (mg/l)	
SICOD = −1.33x + 99.1	$x \leq 20$
SICOD = 103* exp.(−0.0157x) – 0.04x	$x > 20$
Subindex NH_3-N **(mg/l)**	
SIAN = 100.5 – 105x	$x \leq 0.3$
SIAN = 94 * exp.(−0.573x) – 5 * 1 x – 2 1	$0.3 < x < 4$
SIAN = 0	$x \geq 4$
Subindex SS (mg/l)	
SISS = 97.5 *exp.(−0.00676x) + 0.05x	$x \leq 100$
SISS = 71 *exp.(−0.0061x) - 0.015x	$100 < x < 1000$
SISS = 0	$x \geq 1000$
Subindex pH (mg/l)	
SIpH = 17.2–17.2x + 5.02x2	$x < 5.5$
SIpH = −242 + 95.5x – 6.67x2	$5.5 \leq x < 7$
SIpH = −181 + 82.4x – 6.05x2	$7 \leq x < 8.75$
SIpH = 536 – 77.0x + 2.76x2	$x \geq 8.75$

Table 6.
Water quality calculation formula based on WQI [20].

(NEM). **Figure 2** shows the level of water salinity of the groundwater solution for each sampling station. The salinity difference at each station has a natural factor in the soil cavity during the infiltration process occurring in the area. Observations were made on the total solubility of salts or groundwater salinity in the study area given the position of the observation station are located near the coastlines. It is clear that the salinity value at Station 1 (S1) was at a high position throughout three observations compared to other stations. The values obtained at S1 during the SWM were 1.1% and during the MT (1.09%) and during the NEM (1.17%).

Station 2 (S2) recorded the second highest reading among all observation stations. At S2 during the SWM, it was 0.68% while during the MT it recorded 0.6% and during the NEM it increased to 1.24%. Station 3 (S3) and Station 4 (S4) showed a flat average value between the two stations with the value obtained at S3 during the SWM and the MT was 0.48% and decreased during the NEM to 0.44%. While the readings recorded at Station 4 (S4) during the SWM was 0.48% and increased during the MT to 0.52% and during the NEM obtained a value of 0.51%.

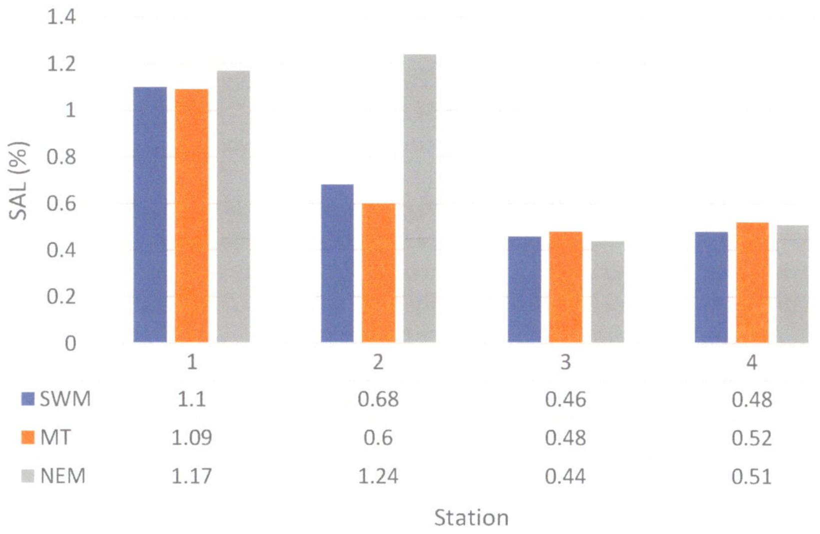

Figure 2.
Value of SAL parameter by station and season.

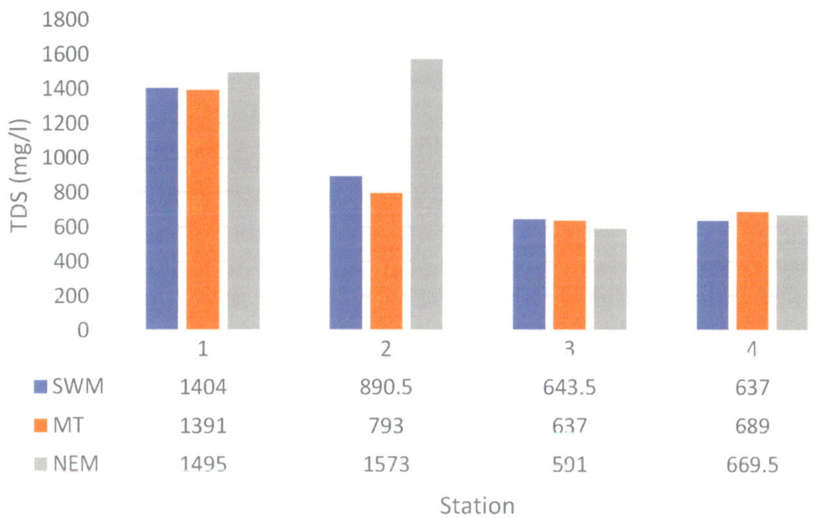

Figure 3.
Value of TDS parameter by station and season.

In the essence, S1 recorded the highest reading of all three observations. This condition may result from seawater penetration into the aquifer system at S1. As pointed by [21], groundwater salinity can also be produced when the boundary between seawater and freshwater moves towards the land due to the leakage of saltwater zones found at the bottom of the aquifer that moves through the joints, fractures or faults. In addition, the locations of S1 and S2 were the closest to the coastlines and this influenced the high SAL value in the area. Overall, the SAL value at all stations exceeded Class I and II but did not exceed Class IV.

Meanwhile, **Figure 3** shows the value of TDS parameters obtained by station and observation by monsoon season in Kapas Island. These different value conditions were also influenced by the SAL content dissolved in groundwater and affected the TDS reading. This study found that the TDS value was in the range of 1573 mg/l and 591 mg/l. Observations showed that the TDS value in the groundwater of the study area was high. NWQSM sets that the TDS for Class II water quality is 1000 mg/l and Class I is 500 mg/l. From the data obtained, the TDS value at S1 was high in all three observations compared to other stations. The values obtained in S1 during the SWM were 1404 mg/l, MT (1391 mg/l) and NEM (1495 mg/l).

Next, S2 recorded the highest reading on the third observation among all other observation stations. At S2 during the SWM, it was only 890.5 mg/l while during the MT it recorded a value of 793 mg/l and during the NEM, it increased to 1573 mg/l. However, S3 and S4 showed a flat horizontal value between the two stations with a small difference in values where S3 during the SWM recorded 643.5 mg/l and during the MT with 637 mg/l and showed a lower value during the NEM with 591 mg/l. While the values obtained in S4 during the SWM (637 mg/l), MT (689 mg/l) and decreased during the NEM (669.5 mg/l).

From the TDS analysis, the TDS was found in abundance at S1 and also showed significant changes at S2 during the NEM. This may be due to the presence of a mineral substance solution dissolved in the water as it is also influenced by SAL which co-exists in the groundwater. The SAL content analysis also increased at S1 and S2. According to Siti Fazilatul Husni et al. [22], this may be due to the rainfall phenomenon which had dissolved more soluble solids while transporting excess sediment and solutions found in the water. Overall, the TDS value at all stations exceeded Class I and II but did not exceed Class IV.

The following discussion is related to the parameters used in determining the WQI for the observation stations. **Figure 4** shows the values of the DO parameter by station and a comparison of different reading values recorded during different monsoon seasons. The study showed that the range of DO content was between 4.68 mg/l to 8.21 mg/l (**Figure 4**). Based on the observations between the stations for the three monsoons showed that there was a change in the DO value between the wet season and the dry season. In fact, the value reading was also constantly increasing from S1 to S4. Starting from the SWM, the analysis obtained showed that the DO value at station S1 which is 7.15 mg/l increased to 8 mg/l at S4. While the value at S2 was 6.77 mg/l and at S3 it was 7.66 mg/l. For observations during the MT, it was found that the DO value at S1 was 6.14 mg/l, increasing to 6.93 mg/l at S2. However, at S3 the DO value decreased to 6.44 mg/l and increased again at S4 which is the highest reading among the stations during the MT with a value of 7.1 mg/l.

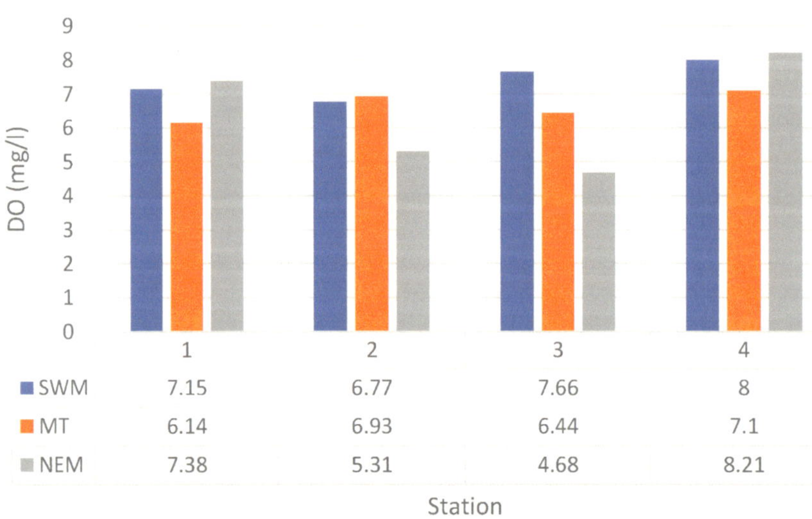

Figure 4.
Value of DO parameter by station and season.

Observations conducted during the NEM showed high-value readings at S1 (7.38 mg/l) and S4 (8.21 mg/l). However, the readings of the DO content decreased at S2 (5.31 mg/l) and decreased to 4.68 mg/l at S3. A low level of DO content can clearly be seen during the NEM at S3 with a value of 4.68 mg/l while the highest DO content was recorded during the same monsoon at S4 which is 8.21 mg/l as there was a heavy rainfall when the observation was made during the NEM. Based on the DO value recorded at all stations, it was found to be in Class I and II except at S3 with a DO value of 4.68 mg/l which is in Class III. It is proven that the rate of DO is dependent on the presence of organic wastes or organic matters that require oxygen causing the level of DO content of the water to be low and it is also influenced by the strength of water convection [23].

Next, the values of the pH parameter were 7.15 to 7.85 (**Table 5**). The results showed that the pH values recorded at all of the observation stations were in Class I as according to the DOE's WQI classification. From the pH values recorded, it showed that the pH of groundwater was in a stable condition for each station according to the monsoon season. High pH readings were recorded at S4 (7.67) during the SWM, NEM (7.61) and MT (7.5). The second highest value recorded was at S2 (7.69) during the NEM and increased to 7.85 during the MT and decreased during the NEM to 7.15. For S3, the MT was high at 7.74 and followed by the value during the SWM (7.56) and decreased during the NEM (7.27). The pH value of 7 obtained showed that the reading is neutral, therefore, the water is neither acidic nor alkaline [24]. While S1 recorded the lowest pH value among all stations during the SWM (7.45), followed by MT (7.43) and the lowest value was during the NEM (7.22). This fluctuation in pH value may be due to the environmental factors of the island as well as the influence of seawater with a higher pH as compared to the pH of freshwater (**Figure 5**) [25].

Figure 6 showed the values of the BOD parameter in Kapas Island that were within the range of 0.38 mg/l to 3.66 mg/l, which was in Class I to III based on the DOE's WQI classification. From the analysis, the differences in BOD readings at all sampling stations during the NEM were the highest compared to other seasons. The data obtained during the NEM at S1 which recorded the highest value of 3.66 mg/l

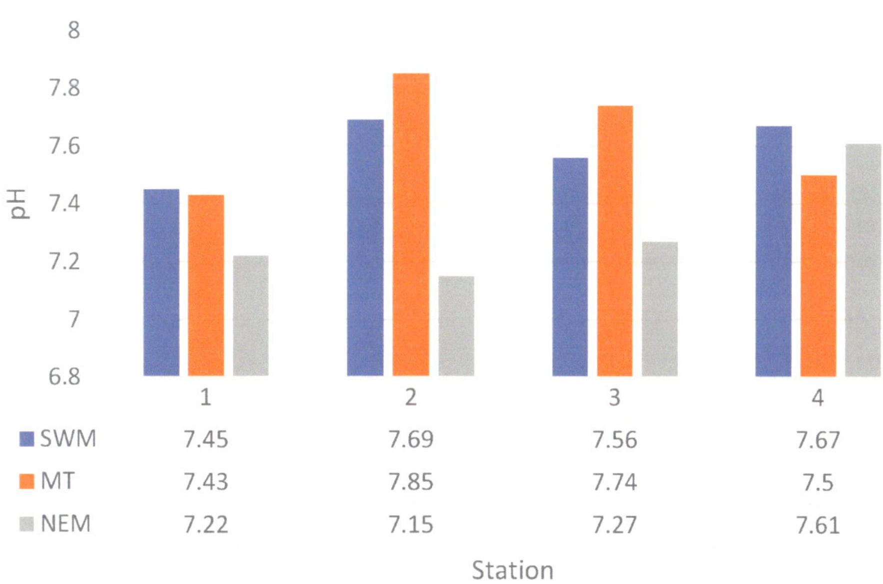

Figure 5.
Value of pH parameter by station and season

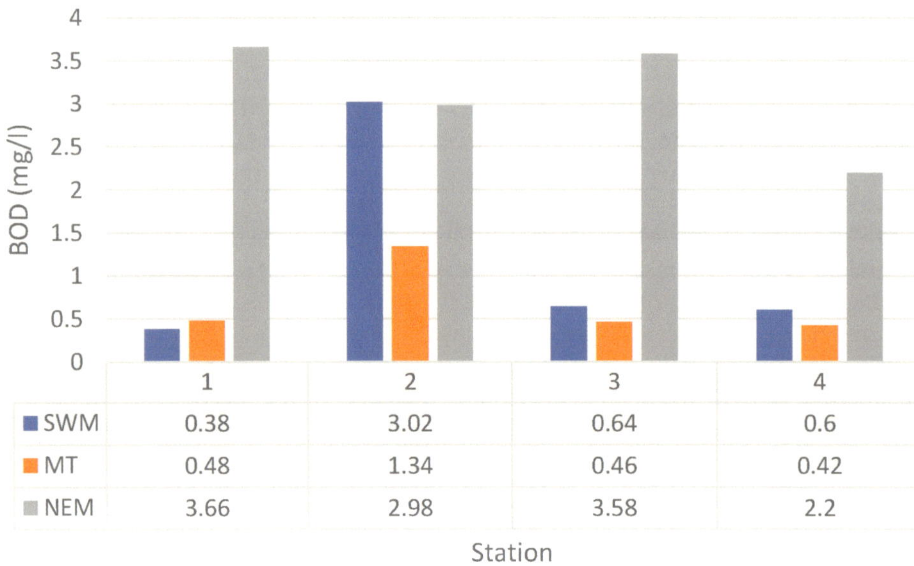

Figure 6.
Value of BOD parameter by station and season.

followed by S3 (3.58 mg/l), S2 (2.98 mg/l) and S4 (2.2 mg/l). During the SWM, all observation stations showed different values and the highest reading of all stations were recorded at S2 with 3.02 mg/l. Meanwhile, the BOD value was recorded as low at S3 (0.64 mg/l), S4 (0.6 mg/l) and S1 (0.38 mg/l). While during the MT, the highest reading values were recorded at S2 (1.34 mg/l) and S1 (0.48 mg/l). The BOD value showed low readings at S3 (0.46 mg/l) and at S4 (0.42 mg/l).

All of the observation stations comparatively showed a high average of BOD values during the NEM. These high BOD values give the impression that there were organic matters that could be broken down by microorganisms and more oxygen was being used because this decomposition process requires oxygen. The high BOD values at all of the stations during the NEM give the impression that the water quality was in Class III based on the DOE's WQI classification. Therefore, the water quality in the area requires further treatment before it can be used and according to Nurfadzlina et al. [17], BOD parameter is used as an indicator for the degree of water pollution. High BOD values indicate that the water tested was in a contaminated status.

Meanwhile, **Figure** 7 showed the COD value according to observation station by season. It was found that the COD concentration of the groundwater in the study area appeared to be not polluted with the majority of the analysis showed not-detected value (nd value) or below the WQI classification. During the study period, the values of COD for all observation stations was within the range of 0 mg/l to 2 mg/l (**Figure** 7). Water quality samples for all stations and seasons were in Class I i.e., below 10 mg/l. Class I indicate that the water analyzed is free from organic pollutants especially from sewage.

The COD values obtained according to stations during the SWM at S2 and S3 were not detected with a reading of nd values followed by S1 (0 mg/l) and S4 (1 mg/l). During the MT, all stations showed that the groundwater is currently not contaminated with an nd reading of the analysis. Next, the COD value during the NEM recorded the highest value at S1 which is 2 mg/l followed by S4 (0 mg/l). While at S2 and S3 readings were obtained with an nd status for both stations. With that value, the COD parameter is in Class I where the quality of groundwater in

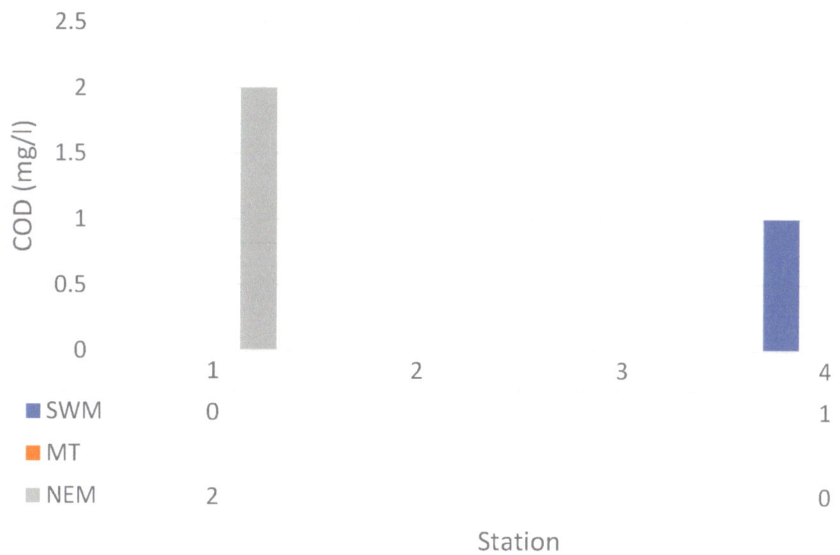

Figure 7.
Value of COD parameter by station and season.

Kapas Island showed no organic pollutants that exist from domestic sewage from residents or tourists. This is due to the widespread usage of COD to determine the concentration of sewage wastes and is used mainly for a mixture of pollutants such as domestic, industrial and biological sewage [25].

Figure 8 shows the readings of the NH_3-N parameter with the NH_3-N concentration for each observation station ranging from 0.01 mg/l to 3.5 mg/l. The majority of the sampling taken and analyzed obtained Class I which is a reading of <0.1 mg/l and only one outlier sample recorded Class V with a reading of 3.5 mg/l as based on the DOE's WQI classification. NH_3-N parameter usually indicates that a body of water has been contaminated by domestic sewage waste, that is, human feces [5, 9].

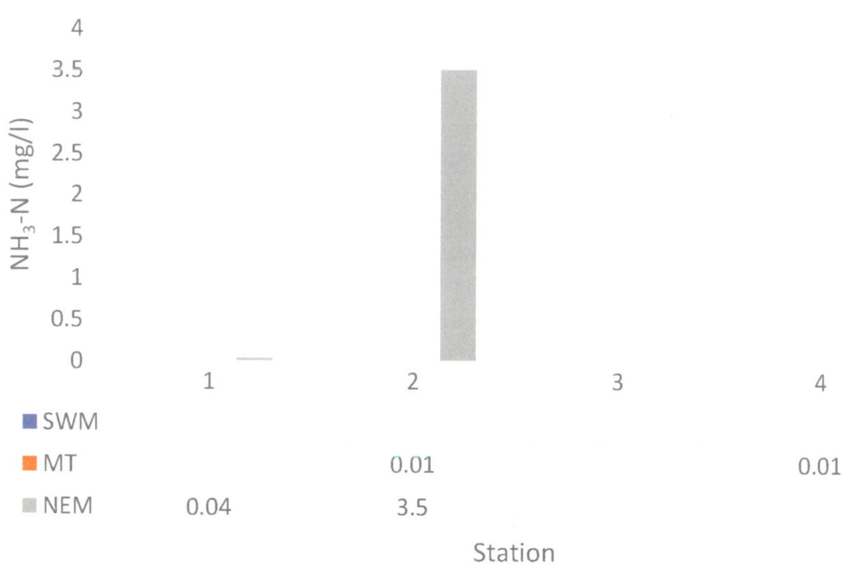

Figure 8.
Value of NH_3-N parameter by station and season.

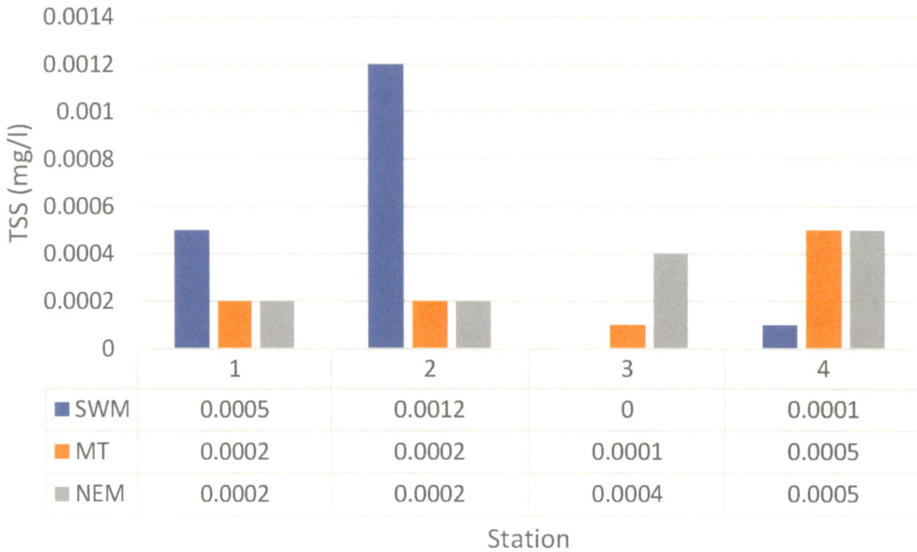

Figure 9.
Value of TSS parameter by station and season.

It is possible that at S2, groundwater had been polluted as a consequence of tourism activities which is sewage waste that was not well managed causing the NH_3-N value to be high in the observations made during the NEM. Nevertheless, the analysis of the study found that the majority of the sample obtained an nd status. During the SWM observation, all stations from S2 to S4 obtained an nd reading while during the MT at S1 and S3 as well as during the NEM at S3 and S4.

Figure 9 showed the laboratory analysis on TSS and found that the values of TSS were within the range of 0–0.0012 mg/l. Based on the DOE's WQI classification, the TSS values for all stations were in Class I which is <25 mg/l. The maximum value recorded was during the SWM at S2 (0.0012 mg/l), followed by S1 (0.0005 mg/l) while at S4 (0.0001 mg/l) and at S3 no TSS was recorded which is 0 mg/l. TSS values were also recorded during the NEM with readings of 0.0005 mg/l at S4 followed by S3 (0.0004 mg/l), S1 (0.0002 mg/l) and S2 (0.0001 mg/l). The average value during the MT was recorded at 0.0005 mg/l at S4 followed by the same values recorded at S1 and S2 which is 0.0002 mg/l as well as at S3 (0.0001 mg/l). The results of this analysis found that the sample obtained through the observations analyzed was water that was not contaminated and also contained less suspended solids.

Furthermore, this situation may be due to the groundwater taken had already gone through suspended solid filtration treatment before the water was being channeled to the tourist accommodation area. The results of the TSS test done in the laboratory may also come from fragments of suspended particles that exist or the condition of contaminated pipeline in the study area because the suspended matter in the water contained inorganic material or organic particles or water-insoluble. Inorganic solids are like clay, silt and other particles in the soil while organic matter includes plant fibers and microorganisms such as algae and bacteria [17].

3.2 Determination of groundwater quality based on DOE's WQI Malaysia

WQI is a very important indicator in knowing the quality status of surface or groundwater. This is carried out to ensure the suitability of water for various purposes such as domestic water supply sources, industrial activities and irrigation in agricultural areas. In this study, DOE's WQI Classification and Water Classes and Uses have been used in determining the status of groundwater quality in Kapas Island. Determination of groundwater WQI in Kapas Island used only six parameters, namely DO, pH, BOD, COD, TSS and NH_3-N. These values of WQI reading are the result of each parameter analyzed based on the WQI formula as shown in **Table 6**. The findings of WQI analysis provided indicators about the current status of groundwater in the study area whether it belongs to Class I (> 92.7), Class II (76.5–92.7), Class III (51.9–76.5), Class IV (31.0–51.9) or Class V (<31.0). Next, based on DOE's Water Quality Classification, groundwater quality status is categorized into three, i.e. clean (81–100), slightly polluted (60–80) and polluted (0–59).

On average, the range of WQI was between 61 to 77, with the majority fell under Class II and III with slightly polluted status (**Table 7** and **Figure 10**). During the SWM, S1 and S3 recorded Class II and S2 and S4 recorded Class III. During the MT, three stations recorded Class II of water quality, namely S1, S3 and S4. Meanwhile, during the NEM, all stations recorded Class III of groundwater quality. Therefore, to be used as a supply of drinking water and other domestic uses, groundwater in this area needs to be treated using conventional and intensive methods. This is for the purpose of providing consumers, especially tourists, with clean water resources. This study is in line with previous studies where an area, especially

Station	SWM			MT			NEM		
	WQI	Class	Status	WQI	Class	Status	WQI	Class	Status
1	77	II	Slightly polluted	77	II	Slightly polluted	73	III	Slightly polluted
2	75	III	Slightly polluted	76	III	Slightly polluted	61	III	Slightly polluted
3	77	II	Slightly polluted	77	II	Slightly polluted	75	III	Slightly polluted
4	76	III	Slightly polluted	77	II	Slightly polluted	76	III	Slightly polluted

Table 7.
WQI values and classes for each groundwater quality observation station in Kapas Island.

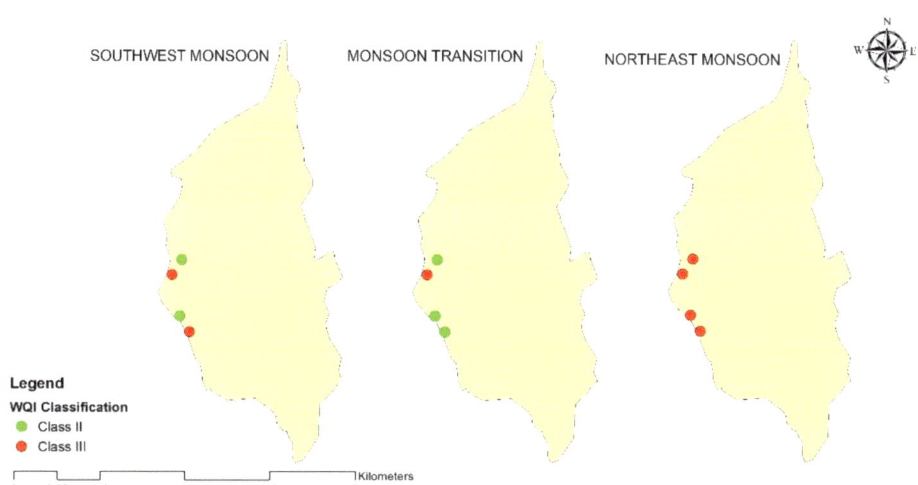

Figure 10.
WQI for observation stations in Kapas Island by season.

in the island area developed for various activities has contributed to the deterioration of groundwater quality [5, 8–9, 13]. Therefore, the values obtained during the NEM for all observation stations in the study area showed relatively the lowest WQI values compared to MT and SWM. This is due to the high concentrations of BOD, COD, TSS and NH_3-N while the low DO value contributed significantly to the deterioration of water quality at all sampling stations taken during the NEM.

4. Conclusion

Groundwater is essential especially in island areas that lack surface water resources. This study which has been conducted in Kapas Island, Marang Terengganu have shown that the groundwater quality was in Class II and III of the slightly polluted status. With the classification obtained, the groundwater in Kapas Island requires conventional treatment for Class II and intensive treatment for Class III. This is because the use of groundwater is not recommended to be directly consumed as drinking water because it contains contaminants from the underground aquifer system during the water pumping process. However, it is still suitable for uses that involve body contact as well as for animal drinks without the need for prior treatment. Therefore, accommodation centers that channel groundwater to be used in tourist accommodation areas should emphasize its use, especially in the process of providing food and beverages to visitors to avoid them from getting sick due to the groundwater.

The determination of WQI status was assessed based on six main parameters that include DO, COD, BOD, SS, NH_3-N and pH tests. Out of six parameters, the parameter with the worst pollution of groundwater quality was the NH_3-N parameter because it exceeded the standard set. The NH_3-N value obtained above this standard was 3.5 mg/l at S2 during the NEM which is in Class V with a highly polluted status. The DO, pH, COD, TSS parameters showed a safe class which is Class I. All stations for DO, pH, COD and TSS parameter tests were below the set standard with an average value of > 7 mg/l for DO, > 7 for pH, < 10 mg/l for COD and < 25 mg/l for TSS. The classification of the obtained parameters was in Class I, so it can be concluded that the groundwater quality in Kapas Island is free from pollution that is for the DO, COD and TSS parameters.

Based on the groundwater quality analysis, it was explained that the status of groundwater quality was safe for external use that included body contact such as bathing, washing, and so on. However, if it is to be used as drinking water and for food preparation, it should be first be given conventional and intensive treatment before use. The lowest WQI value obtained for groundwater quality in Kapas Island was 61%. This value has recorded a Class III of moderately contaminated status but is still safe for external body use and also safe to be used as a drinking source for animals.

Author details

Mohmadisa Hashim[1*], Arijatul Wardah Ahmad[1], Zahid Mat Said[2], Nasir Nayan[1], Hanifah Mahat[1], Yazid Saleh[1] and Koh Liew See[1]

1 Department of Geography and Environment, Faculty of Human Sciences, Sultan Idris Education University, Tanjong Malim, Perak, Malaysia

2 Department of Biology, Faculty of Sciences and Mathematics, Sultan Idris Education University, Tanjong Malim, Perak, Malaysia

*Address all correspondence to: mohmadisa@fsk.upsi.edu.my

References

[1] Fadaei A, Sadeghi M. Evaluation and assessment of drinking water quality in Shahrekord, Iran. Resources and Environment. 2014; 4(3): 168-172.

[2] Krishan G, Singh S, Singh RP, Ghosh NC, Khanna A. Water quality index of groundwater in Haridwar District, Uttarakhand, India. Water and Energy International. 2016; January 2016: 55-58.

[3] Kumar SK, Logeshkumaran A, Magesh NS, Godson PS, Chandrasekar N. Hydro-geochemistry and application of water quality index (WQI) for groundwater quality assessment, Anna Nagar, part of Chennai City, Tamil Nadu, India. Appl. Water Sci. 2015; 5: 335-343.

[4] Luthfi A, Maryunani SM, Cahyo H. Analysis of sustainable groundwater resources management in Jember District (Study in Sumberjati Village, Silo Sub-District). Resources and Environment. 2017; 7(5): 115-123.

[5] Mohamed AAJ, Abdul Rahman I, Lim LH. Groundwater quality assessment in the urban-west region of Zanzibar Island. Environ. Monit. Assess. 2014; 186: 6287-6300.

[6] Rahayu EO, Aji A, Tukidi. Sebaran kelayakan air sumur sebagai air baku minum di Kelurahan Sekaran Kecamatan Gunungpati Kota Semarang tahun 2018. Geo Image. 2019; 8(1): 1-8.

[7] Khan A, Qureshi FR. Groundwater quality assessment through Water Quality Index (WQI) in New Karachi Town, Karachi, Pakistan. Asian Journal of Water, Environment and Pollution. 2018; 15(1): 41-46.

[8] Koh DC, Ko KS, Kim Y, Lee SG, Chang HW. Effect of agricultural land use on the chemistry of groundwater from basaltic aquifers, Jeju Island, South Korea. Hydrogeology Journal. 2007; 15: 727-743.

[9] Lin CY, Abdullah MH, Musta B, Aris AZ, Praveena SM. Assessment of selected chemical and microbial parameters in groundwater of Pulau Tiga, Sabah, Malaysia. Sains Malaysiana. 2010; 39(3): 337-345.

[10] Moayedi H, Huat BBK, Asadi A, Kazemian S, Mohammad Ali TA. Groundwater quality assessment of Labuan Island. International Journal of the Physical Sciences. 2011; 6(18): 4441-4449.

[11] Aris AZ, Abdullah MH, Praveena SM. Evolution of groundwater chemistry in the shallow aquifer of a small tropical Island in Sabah, Malaysia. Sains Malaysiana. 2009; 38(6): 805-812.

[12] Mohd Zul Asyraf MK, Md Hashim N, Mohd Zin MS. Seawater intrusion in the ground water aquifer in the Kota Bharu District, Kelantan. PEOPLE: International Journal of Social Sciences. 2015; Special Issue 2015: 320-327.

[13] Mondal NC, Singh VS, Saxena VK, Prasad, RK. Improvement of groundwater quality due to fresh water ingress in Potharlanka Island, Krishna delta, India. Environ. Geol. 2008; 55: 595-603.

[14] International Groundwater Resources Assessment Centre. Groundwater monitoring in small island developing states in the Pacific. Netherland: IGRAC; 2016.

[15] Theophile R, Razanamparany B, Andrianainarivelo M. Physico-chemical analysis of the water of the well of Ampombilava in the District of Nosy-Be. Resources and Environment. 2017; 7(3): 81-86.

[16] Marang District Council. Destinasi Menarik (Internet]. 2018. Available from http://www.mdmarang.gov.my/ms/pelawat/destinasi-menarik/pulau-kapas [Accessed: 2020-10-13]

[17] Nurfadzlina MR, Munirah DS. (2012). Kajian kualiti air sungai Bedil berdasarkan "Water Quality Index (WQI). 2012. In: Seminar Penyelidikan & Inovasi 2012 Giatmara Bandar Kuching. Available from https://docplayer.net/47781054-Kajian-kualiti-air-sungai-bedil-berdasarkan-water-quality-index-wqi.html [Accessed: 2020-11-10]

[18] Muhammad Fuad R, Wan Syaidatul Aqma WN, Aznan F, Norsyafina R. Analisis kualiti air bawah tanah di Lebuhraya SILK Kajang – Sungai Long, Hulu Langat, Selangor. Undergraduate Research Journal for Earth Sciences. 2017; 2017: 46-50.

[19] American Public Health Association. Standard methods for the examination of water & wastewater. (Vol. 21). United States: American Public Health Association; 2005.

[20] Department of Environment. Environmental quality report 2017. Putrajaya: Department of Environment, Ministry of Natural Resources & Environment; 2018.

[21] Umar H, Abdul Rahim S, Abdul Ghani R, Khairul Azlan R. Kemasinan air perigi dan subpermukaan lembangan tuba Langkawi dengan analisis hidrokimia dan survei keberintangan elektrik menegak. Sains Malaysiana. 2009; 38(6): 851-856.

[22] Siti Fazilatul Husni MS, Wan Ruslan I, Sumayyah Aimi MN. Kualiti air dan ciri fizikokimia lembangan sungai input ke dalam takungan kolam Bukit Merah, Perak. In: Mohamad Suhaily Yusri Che Ngah, Mohmadisa Hashim, Nasir Nayan, editors, Hidrologi dan pengurusan sumber air di Malaysia. Tanjong Malim: Department of Geography & Environment; 2013. p. 67-84.

[23] Noraini M, Mohd Adam O, Sumayyah Aimi MN. Kualiti air di empangan Timah Tasoh dan kawasan tadahannya. In: Mohamad Suhaily Yusri Che Ngah, Mohmadisa Hashim, Nasir Nayan, editors. Hidrologi dan pengurusan sumber air di Malaysia. Tanjong Malim: Department of Geography & Environment; 2013. p. 99-113.

[24] Rosner B. Measuring the "Power of hydrogen ions" (pH). 2018. Available from https://www.carolina.com/teacher-resources/Interactive/measuring-the-power-of-hydrogen-ions/tr10793.tr [Accessed: 2020-7-5]

[25] Suratman S, Norhayati MT. Kesan antropogenik terhadap kualiti air di lembangan sungai Marang, perairan Laut China Selatan. Sains Malaysiana. 2013; 42(6): 743-751.

Chapter 4

Analysis of Ground Water from Selected Sources in Jalingo Metropolis, Nigeria

Benjamin Ezekeil Bwadi, Mohammed Bakoji Yusuf, Ibrahim Abdullahi, Clement Yakubu Giwa and Grace Audu

Abstract

Water is very significant in the development of a stable community, but many societies are confronted with the challenges of poor wastes management system with indiscriminate waste disposal and bad land practices, which easily pollute water sources and consequently degrade water quality. This study was to analyze the physicochemical properties of ground water from multiple point sources in Jalingo, Taraba state of Nigeria. Water samples were collected from twenty seven (27) sites from the study area during the raining and dry seasons. The analysis was carried out to determining the physico-chemical properties of the ground water and comparing with the World Health Organization (WHO) standard for drinking water. The physicochemical properties of ground water analyzed include; odor, taste, temperature and electrical conductivity were tested in the field using water meter tester. Whereas pH, total dissolved solids, alkalinity, hardness, salinity, iron, manganese, fluoride, nitrate, nitrite, chloride, sulphate and dissolved oxygen were analyzed in the laboratory using Wagtech potable water testing equipment. The physical properties of water analyzed were temperature, odor, taste, and turbidity. Whereas the chemical properties of water analyzed were pH, electrical conductivity (EC), total dissolved solids (TDS), alkalinity, hardness, salinity, iron (Fe), manganese (Mn), fluoride (F^-), nitrate (NO_3^-), nitrite (NO_2^-), Chloride (Cl^-), sulphate (SO_4^{2-}), dissolved oxygen (DO).The result shows the range of the mean values of the temperature(26.7–33.1) °C, p H(6.5–8.9), Fe (0.01–0.08 mg/L), NO_3^- (0.01–38.5 mg/l), NO_2^- (0.01–0.09 mg/l), Mn (0.01–0.17 mg/l), F(0.01–0.82 mg/l), alkalinity(39-204 mg/l), salinity (42-508 mg/l), SO_4(14-93 mg/l), total dissolved solids (6–637) mg/l, turbidity(0.4–10.6 mg/l), hardness(48-187 mg/l), and fecal coliforms(1–4)fcu/100mi, dissolved oxygen(1.1–6.87)mg/l, EC(10.99–1066) ohm/cm, Cl (10-320 mg/l). All except alkalinity and hardness are within the WHO permissible standards of quality drinking water. The highest alkalinity (204 mg/l), hardness (187 mg/l) and low dissolved oxygen (6.87 mg/l) attributed to the high concentration of dissolved salts and basic cations in the water. The methodology applied in the study was effective in analyzing the physicochemical properties of water in the study area. Therefore, it was recommended that there should be frequent water source testing by stakeholder in water resources with the view to treating the water. Policy maker should also enforce the regulation of the use of chemical fertilizers, agro-chemicals and the indiscriminate waste disposal.

 IntechOpen

Keywords: ground water, physicochemical properties, water quality, multiple sources

1. Introduction

Water plays a vital role in the development of a stable community and society, since human being can exist for days without food, but absence of water for a few days may lead to death [1]. The essential nature of water to man's daily usage vis-à-vis quantity and quality right from time immemorial has been on the increase [2]. Unfortunately, drinking water in developing countries especially in Nigeria in particular is susceptible to toxins as a result of effluents and pollutants [3, 4].

Water is a common natural chemical substance containing two atoms of Hydrogen and an atom of Oxygen. Its common usage refers to liquid form, though has other forms: solid water- ice and gaseous forms - water vapor and steam. Water is indispensable for life and socioeconomic development of any society. It is used in domestic activities (cooking, drinking, washing, bathing etc.), agricultural activities (e.g. irrigation, gardening), generation of power (hydroelectric power plants), running industries, recreational activities etc. It is very essential for human existence and sustenance of life. Water constitutes 60–70% of the total body weight. A man can live for several days without food but will only survive for few days without water. Therefore, water is indispensable for normal physiological function of plants and animals (Guyton, 1996 cited in [5]). In spite of its importance in sustenance of live and livelihood, it is the major cause of morbidity and mortality because of limitations in access and quality [5]. The basic physiological requirement for drinking water has been estimated at about 2 liters per capita per day which is just enough for survival [6]. World Health Organization (WHO) states that domestic water consumption of 30–35 liters per capita per day is the minimum requirement for maintaining good health. However, the amount of water required by individuals varies depending on climate, standard of living, habit of the people and even age and sex.

One factor that impinges more on the accessibility to enough quality drinking water is the distance of the source from house. This condition forces the individual most especially the women and children (especially girls) to transverse many kilometers to get safe drinking water (which deprives them from engaging in productive ventures or going to school like their male counterparts). In addition to this, in order to reduce the hardship in getting water, they may resort to reducing the quantity of water used in the house far below the recommended volume and also they may resort to fetching water from unimproved sources e.g. unprotected well, pond, stream etc. [7].

Water quality refers to the chemical, physical and biological characteristics of water [8]. It is a measure of the conditions of water relative to the requirement of one or more biotic species, and or to any human need or purpose. It is most frequently used by reference to a set of standards against which compliance can be assessed. The most common standards used to access water quality relate to health of ecosystems, safety of human contact and consumption [9].

Safe (quality) drinking water is that which does not present any significant health risk over a life time consumption, including any sensitivities that may occur in different stages of life [10]. It is water which is free from pathogenic microbes, hazardous chemicals/substance and esthetically acceptable (i.e. pleasing to sight, odorless and good taste). It is important that this type of water should not only be available, but also be available in enough quantity all the time, i.e. twenty-four hours a day, seven days a week ("24/7").

In assessing quality of drinking water, physical, chemical and bacteriological parameters must be considered. Although water from a source may not pose any health threat to consumers, they may abhor it due to its color, odor, or taste [10]. Physical parameters include color, smell, temperature, pH, turbidity etc. There are myriad of chemical substances which may be naturally present or introduced (even chemicals used for water treatment) into water; those that are naturally present seldom pose risk to health. However, chemicals released due to anthropogenic activities (fertilizer, pesticides, herbicides, industrial effluents and byproducts etc.) carry more health risk to consumers.

Fortunately, whether chemical naturally present or introduced into water, there are maximum allowable concentration (limit) of most of them proposed by World Health Organization (WHO), which serves as guide. Some of the chemical substances include residual chlorine (RC), Iron (Fe), Fluoride (Fl), Nitrate/Nitrite, Lead (Pb), Mercury (Hg) [10, 11].

Bacteriological (microbial) parameter is used to assess drinking water quality using the index /indicator concept as advocated by Waite (1991, cited in [5]). The infectious risks associated with drinking water are primarily those posed by fecal pollution and their control depends on being able to assess the risk from any water source and applying suitable treatment to eliminate the risk. Rather than trying to detect the presence of pathogens at which time the consumer is being exposed to possible infection, it is better practice to look for organisms, while not pathogenic themselves, that show the presence of fecal pollution and therefore the potential for the presence of pathogens. For this reason, *E. coli* (*E.coli*) is universally used as an indicator organism to assess water treatment and widely preferred as index organism for fecal contamination. Thermo-tolerant coliform count (Fecal coliform) is acceptable where *E. coli* detection is not possible [10].

The presence of other microbes may indicate fecal contamination as well e.g. fecal streptococci indicate recent contamination of water sources with feces. While planet earth is made up of predominantly water, only 3% is fresh water, and of this, 99% is trapped in icecaps and glaciers. Even the 1% of the fresh water available for human use is not evenly distributed [11]. According to WHO and UNICEF report [7], safe water which is a basic necessity is still a luxury for many poor developing countries of the world today. It has been estimated that over 1.1 billion people do not have access to drinking water from improved sources. Eighty percent of the unserved populations live in these three regions – Sub-Saharan Africa, Eastern Asia and Southern Asia. Eighty- four percent of these people are the rural dwellers.

In 2004, 83% of the world population (5.3 billion) had access to drinking water from improved sources. This seemingly high global statistics hide a critical situation in some developing countries. In sub-Saharan Africa and Oceania only 54% and 50% of their populations respectively are served with improved sources of drinking water in 2004. Whereas at the same period, the population that had access to water from improved sources is over 90% in the Caribbean, Northern Africa and Western Asia [12].

Although over 80% of the developing world population has access to some type of improved drinking water source, only 44% have access through household connection from piped system. There are large disparities between regions. While access to drinking water through household connection is as low as 16% in Sub-Saharan Africa, 20% in South-Eastern and 21% in Oceania; it is much higher in Eastern Asia 70%, Northern Africa 76%, and 80% in Latin America and the Caribbean [12].

Jalingo metropolis, which is located in sub-Saharan Africa, has no official record of households' connection to water, but with the heavy dependence on ground water, it is evident that there is little or no connection to improved drinking water within the study area. This necessitated the current study on physicochemical properties of ground water from multiple sources in Jalingo metropolis.

Jalingo as an urban centre is confronted with the challenges of poor waste management system and fast urban sprawl that is evident within the township. Indiscriminate waste disposal coupled with bad land practices are common scenes in the metropolis. These unguided practices according to Ndabula & Jidauna [13] and Dabi Jidauna [3] can easily pollute surface water, and consequently degrading of the water quality. The residents of Jalingo Metropolis depend on both surface and ground water sources for their different water uses. The most dependable sources observed include tap water from the water board, water vendors, wells, and6sometimes riverbed are used as sources of water supply. These sources are often prone to pollutants that are categorized as heavy metals that are often associated to human activities and further exacerbated by urban sprawling and poor waste management [4]. Therefore, the need to access a reliable, secure, safe, and sufficient source of fresh water is a fundamental requirement for the survival, well-being, and socio-economic development of all humanity [14]. Hence, the desire and the need for portable water supply cannot be over emphasized. Irrespective of sources, domestic water supply should be water of high quality, while water for other uses can be of moderate quality. Moreover, most people in the urban centers often depend on the water vendors for domestic water supply, and in many cases, the water accessed is used directly without treatment. Drinking water is water of highest quality, while water of good quality can be put into any other use [2].

There has been heavy investment in terms of pipe borne or tap water in order to increase access of residents to quality water for drinking and domestic uses, yet it is no news that the residents of Jalingo have no access to quality drinking water especially as the sources are diverse. Thus, this study analyzed the physicochemical properties of ground water from different sources within the metropolis, hence to create awareness towards developing a strategy for monitoring and protecting groundwater quality of the area.

2. Method and materials

2.1 Study area

The study area is Jalingo metropolis in Taraba state located at latitude 8^0 47′ to $9^0$01′N of the equator and longitude $11^0$09′ to $11^0$30′E of the Greenwich Meridian (**Figure 1**). It is situated within the Northern Guinea Savannah ecological zone. It is bounded to the North by Lau LGA, to the East by Yorro LGA, to the South and West by Ardo-Kola LGA. Jalingo has a total land area of about 195.071km^2. Nine of the ten (10) wards are located within the metropolis. These include Barade, Kona, Maji Dadi, Sarkin Dawaki, Sintali A, Sintali B, Turaki A, Turaki B and Kachalla Sembe wards [15]. Presently, Jalingo has a projected population [16] of 191,443 people at 2.83% annual growth rate [17] (**Figure 2**).

The WHO standards follow these classifications also (**Table 1**).

2.2 Research design

The study adopted the experimental research design. This involves studying phenomenon in a laboratory or natural setting, and hypotheses are amenable to statistical testing because it is possible to observe and measure variables in the investigation.

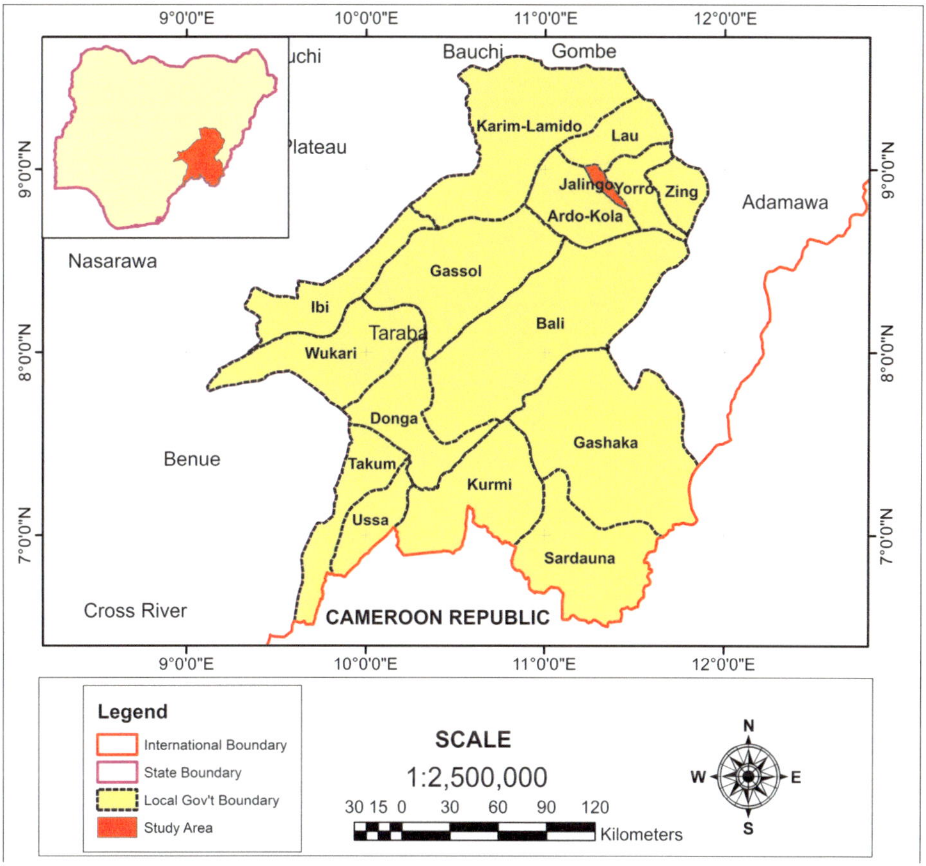

Figure 1.
Map of Taraba showing study area.

2.3 Population of the study

The population of the study comprises all drinking water sources within Jalingo metropolis. Due to the vastness of the metropolis, the researcher, randomly sampled three water sources from each of the nine wards in the metropolis making 27 sampling points.

2.4 Sample size and sampling technique

The sample size for the study is 27 water sources (sampling sites) within Jalingo metropolis. Purposive sampling was used to select the nine (9) wards in Jalingo metropolis. Random sampling was used to select 27 sampling sites, three (3) sampling sites each per ward.

2.5 Sources of data

The primary data was sourced through Ground Truth Observations (GTOs) in the field. This includes collection of water samples from three (3) major locations from each of the nine wards; Sarkin Dawaki(SD), TurakiA(TA), Turaki B(TB), Majidadi(MD), Kachalla Sembe(KS), Kona(KL), Sintali A(SA), Sintali B(SB) and Barade(BR) in Jalingo Metropolis.

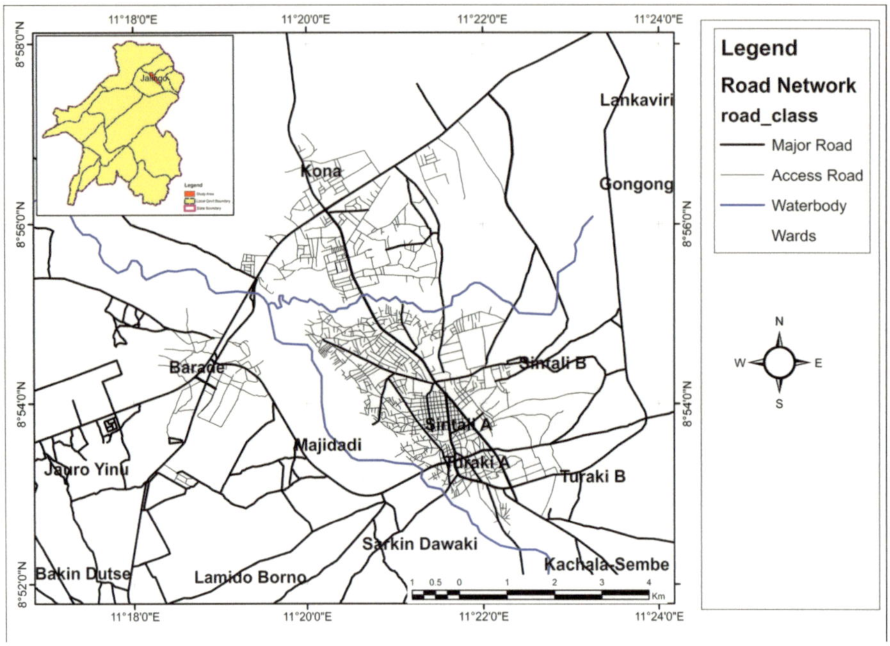

Figure 2.
Administrative map of the study area (min. Of Land & Survey, Jalingo).

S/N	Parameter	Maximum permissible limits in water				
		NAFDAC	SON	FEPA	UNICEF	WHO
1	Conductivity (S/m)	1000	1000	70	1000	0–1500
2	Total dissolved solids (mg/l)	500	500	500	500	1000
3	Hydrogen ion concentration	6.5–8.5	6.5–8.5	6.0–9.0	6.5–8.5	6.8
4	Total hardness (mg/l)	100	100	—	150	100
5	Total alkalinity (mg/l)	100	100	—	200	100
6	Nitrate (mg/l)	10	10	20	50	50
7	Water temperature (^{0}C)	—	—	26	6.5–8.5	40
8	Dissolved oxygen (mg/l)	—	—	≥4		≤6
9	Turbidity (NTU)	—	—	—	15	5 NTU
10	Iron (mg/l)	—	—	—	0.2	0.1–0.5
11	Fluoride (mg/l)	—	—	—	1.5	0.5–1.5
12	Nitrite (mg/l)	—	—	—	0.2	50
13	Chloride (mg/l)	—	—	—	250	250
14	Sulphate (mg/l)	—	—	—	250	100
15	Manganese (mg/l)	—	—	—	0.2	0.1–0.5
16	Salinity (mg/l)	—	—	—	500	≤ 600

Source: Adejuwon and Adelakun [18]. Where: NAFDAC-National Administration for Food, Drugs and Control, SON-Standard Organization of Nigeria, FEPA-Federal Environmental Protection Agency, United Nation International Children's Emergency Fund-UNICEF and WHO-World Health Organization.

Table 1.
Selected national and international water quality standard guidelines.

The secondary data were obtained from journal publications, books, WHO water standards, online articles, published and unpublished theses.

2.6 Data collection and preparation of samples

One-liter capacity sample bottles were used for collection of water samples. The bottles and their covers were washed and properly sterilized. When collecting water samples, the sterilized bottles were rinsed with the water to be sampled three times before finally collecting the samples.

Water samples from taps, well and boreholes were collected using the one-liter sterilized bottles fitted with its cover. The sample bottles were filled with water by a gentle flow, for water samples from taps and boreholes. The bottles were then covered and labeled for easy identification. Water samples from wells were fetched to fill a sterilized plastic bucket and transferred immediately into the 1-liter sample bottles. All water samples were stored in cooler containing ice packs and were transported immediately within some few hours to the laboratory where the analysis was carried out.

In this study, 26 samples of ground water and one sample pipe-borne water were collected. The samples were collected from twenty-seven (27) sites in nine (9) locations (KL, BR, SB, SD, MD, TA, TB, KS and SA) all Jalingo Metropolis. The coordinates of each water source sample site were obtained using handheld GPS. The coordinates were recorded and labeled accordingly. This was the data used for drawing the map of water sampling points (**Figure 3**).

2.7 Analytical procedure

Data analysis involves analytical methods of sorting, scrutinizing, processing and translating data into research findings based on the set objectives of research. Some of thesimple measurements were made on-site like temperature, pH, and conductivity, in direct contact with the water source in question. More complex

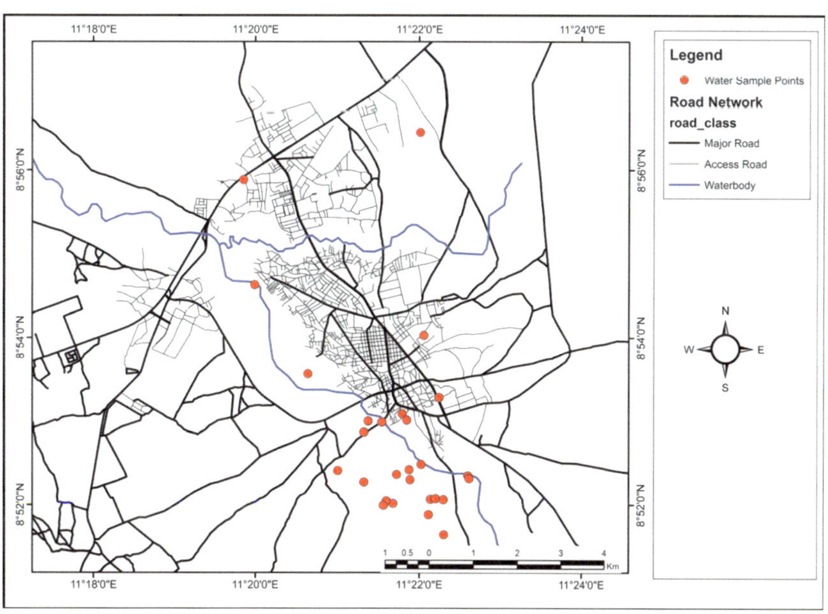

Figure 3.
Map of study area showing sampling points.

measurements were made in a laboratory setting where water samples were collected, preserved, and analyzed in the laboratory. For the purpose of this study, only some properties were examined. These include pH, turbidity, color, odor, hardness, alkalinity, salinity, total dissolved solids (TDS), dissolved oxygen (DO) chloride (Cl), nitrate (NO^3)/nitrite (NO^2), fluoride (F), iron (Fe), sulphate (SO_4^{2-}), manganese (Mn^{2+}), fecal coliform and total fecal coliform.

2.7.1 Physicochemical parameters

pH: Water pH was analyzed using Wagtech WE30200 pH meter. The acceptable pH for drinking Water ranges from 6.5–8.5.

Turbidity: Turbidity was measured using Wagtech WE30140 Potalab Turbidimeter. The turbidity measurement was conducted by placing the meter on a flat surface, filling a clean sample vial to mark, placing in a sample well and covering the vial with light shield cap. The display reading was recorded as sample turbidity. The maximum acceptable value for turbidity of drinking water is 5NTU.

Total Dissolved Solids (TDS) and Electrical Conductivity: were determined using Wagtech WE30120 conductivity/TDS meter.

Chloride: was measured by complexometric titration of 100 ml sample using 0.141moldm^{-3} silver nitrate ($AgNO_3$) in the presence of 1 ml potassium chromate indicator (K_2CrO_4) at pH of 7–8. At the endpoint titration color changes from yellow to pinkish-yellow and the chloride concentration was computed by calculation as follows:

Mg/l Chloride = (A-B) x N x35,450/ vol.of sample.

Where A = Sample titer value, B = Blank titer value and N = 0.0141.

Nitrate/Nitrite: was determined via reduction method and the resulting, nitrite determined by reaction with sulphanilic acid in the presence of N – (1-naphathyl)-ethylene diamine to form reddish dye. The intensity of color produced is directly proportional to the nitrate transmittance obtained was converted to concentration with aid of nitratest calibration chart and mg/l NO_3 obtained by multiplying the result by a factor of 4.4. The Photometer was calibrated with the water sample to be tested.

Fluoride: was analyzed by adding Zirconyl chloride and Eriochrome Cyanine reagents tablets to a 10 ml sample of water in acid solution to form a red colored complex. This is destroyed by fluoride ion to give a pale yellow. The color produced was directly proportional to the fluoride concentration and was measured using Wagtech WE10441 Potalab Photometer 7100 at 570 nm wavelength. Percent transmittance obtained was converted to mg/l F with the aid of fluoride calibration chart. The Photometer was calibrated with the water sample to be tested.

Iron (Fe): was measured by using Wagtech spectrophotometer. The photometer was calibrated with the water sample to be tested. The color produced was directly proportional to the iron concentration and was measured using the Wagtech photometer at wavelength 570 nm. Percent transmittance obtained was converted to mg/l Fe with aid of iron calibration chart.

Sulphate: was determined by modified turbidimetric method with barium in sulpha Ver 4 sulphate reagent. The color produced was directly proportional to the sulphate concentration and was measured using DR/2010 Hach Spectrophotometer at 450 nm wavelength. The spectrophotometer was calibrated with the water sample to be tested.

Manganese: was determined by direct aspiration into an Air- Acetylene Flame using Atomic Absorption Spectrometer.

Total Alkalinity: was measured by titrating 100 ml sample using 0.01moldm^{-3} of H_2SO_4, phenolphthalein indicator, methyl orange indicator and pH meter at endpoint p H OF 4.5 and the Total Alkalinity as mg/l $CaCO_3$ = Titer value/vol. of sample x 1000.

Total Hardness: was measured by complexometric titration of 100 ml sample using 0.01moldm^{-3} disodium slat of ethylene diamine tetra acetic acid (EDTA) in the presence of Eriochrome Black T. at the titration endpoint color changes from wine red to bluish-green and the total hardness content was computed by calculation as follws:

Total Hardness as mg/l $CaCO_3$ = Titer value/vol. of sample x 1000.

Dissolved Oxygen(DO): Was determined using Wagtech potalab spectrophotometer water analysis kit.

Total coliforms: Was determined using the membrane filtration technique using Wagtech field kits.

Temperature: The temperature was determined using the Wagtech field water meter.

Water Taste and Odor: were determined in field by drinking and perceiving the smell and taste of the water at sites.

2.8 Statistical analysis

ANOVA analysis was carried out using statistical software (SPSS version 23.0) to compare the relationship between water sources and the physicochemical characteristics of water in Jaingo Metropolis.

3. Results and discussion

This section presents the results of the analysis of the data collected during the field survey carried out, and the interpretation and discussion of the results. The variables analyzed in the study include the physicochemical properties of the drinking water, comparison of physical and chemical characteristics of drinking water in Jalingo side-by-side WHO standard quality of drinking water in developing countries and the relationship between water sources and the physicochemical characteristics of water in Jaingo Metropolis. In all, the study examined twenty (20) physicochemical parameters of water.

3.1 Temperature

The temperature of water samples from all sites during the dry season range from 26.7°C – 33.1°C as shown in **Table 2** with mean value range of 28.3°C to 31.8°C as shown in **Table 3** and the values of temperature recorded during the rainy season ranged from 21.2°C to 28.5°C as shown in **Table 4** with mean value of 22.35 to 27.37°C as shown in **Table 5**. The high temperature values obtained from sites BR3 (33.1°C), MD3 (32.4°C) and SA3 (32.2°C) during the dry season were higher than the values of 25.87 to 27.56°C reported by Ikhuoriah and Oronsaye [19] from water samples collected from river source. The temperature values obtained from this study are within the WHO permissible limit of 40°C.

3.2 pH

pH is one of the important water quality parameters. Measurement of pH relates to the acidity or alkalinity of the water. A sample is said to be acidic if the pH is below 7.0 and it is said to be alkaline if the pH is higher than 7.0. Acidic water can

Site	Taste	Odor	Temp°C	pH	EC	TDS	TURB	Alkalinity	Hardness	Salinity	DO (mg/1)
KL1	Unob	Unob	30	7.8	10.99	6	3	120	121	42	1.4
KL2	Unob	Unob	27.5	6.5	452	226	6.2	168	182	68	4.8
KL3	Unob	Unob	27.5	8.1	89.5	45	4.8	39	48	42.8	4.3
BR1	Unob	Unob	26.7	7.6	766	383	0.2	168	147	78	1.1
BR2	Unob	Unob	27.9	8.1	573	282	1.2	140	138	42	1.72
BR3	Unob	Unob	33.1	8.4	343	171	1.1	121	110	58.2	1.4
SB1	Unob	Unob	29.3	8.6	320	160	1	110	90	42	1.13
SB2	OBJ	Unob	27.7	8.2	480	240	1.6	123	171	58.2	1.5
SB3	Unob	Unob	30.5	8.4	423	211	1.1	140	187	65	1.2
SD1	Unob	Unob	29.2	8.8	93.5	47	2	84	67	33.4	5.6
SD2	OBJ	Unob	28.1	8.9	921	460	1.9	194	148	482	4.53
SD3	Unob	Unob	30.6	8.9	963	482	1.4	168	171	120	1.83
MD1	Unob	Unob	29.6	8.6	143.4	72	3.6	110	121	48	2.73
MD2	OBJ	Unob	30.8	8.9	915	452	2	182	148	496	1.14
MD3	OBJ	Unob	32.4	8.4	1261	630	1.8	204	138	508	5.8
TA1	Unob	Unob	28.1	8.7	958	480	0.4	161	128	52.7	4.85
TA2	OBJ	Unob	29.1	8.7	1066	533	0.8	198	132	71	5.6
TA3	Unob	Unob	28.1	8.4	637	637	0.7	144	114	64	5.22
TB1	Unob	Unob	30.5	8.8	827	414	10.6	182	148	56	4.63
TB2	OBJ	Unob	30.7	8.2	708	354	2.4	130	186	71	6.87
TB3	OBJ	Unob	30.5	8.6	703	352	1.8	128	126	65	4.4
KS1	OBJ	OBJ	29.4	8.9	685	343	1.4	160	132	74	2.64
KS2	Unob	Unob	29.3	8.6	762	381	1.2	161	128	64	1.7
KS3	Unob	Unob	30.1	7.9	633	316	0.9	142	113	64	3.2
SA1	Unob	Unob	31.7	7.8	465	233	0.9	140	131	58	3.62
SA2	Unob	Unob	31.5	8.4	640	320	1.4	144	118	61	4.61
SA3	Unob	Unob	32.2	8.2	404	202	0.8	160	181	72	3.51

Table 2.
Physicochemical properties of water in Jalingo Metropolis during dry season.

lead to corrosion of metal pipes and plumbing system as well as harmful to human life.

The pH values from this range from (6.5–8.9) as shown in **Table 2** with mean value of 7.5 ± 0.85–8.9 ± 0.06 shown in **Table 3** during the dry season and values of 5.2–6.7 as shown in **Table 4** with mean value of 5.93 ± 0.67–6.6 ± 0.1 as shown in **Table 5** during the rainy season.

The highest pH value recorded during the rainy season was 6.7 which were higher than the range of 5.76 to 6.01 reported by Ikhouriah and Oronsaye [19] in Ossiomoo river ologbo - a tributary of Benin River, Southern Nigeria.

The pH values reported in this study is also higher than the range of 5.96–5.54 reported by Ohinedu, Nwinyi, Oluwadamisi and Eze [20] is assessing the water quality in cananland, Ota, Southwest Nigeria.

DOI: http://dx.doi.org/10.5772/intechopen.99082

Site	Temp°C	pH	EC	TDS	TURB	Alkalinity	Hardness	Salinity	DO (mg/1)
KL	28.3 ± 1.44^{a}	7.5 ± 0.85^{a}	184.2 ± 235.3^{a}	92.3 ± 117.4^{a}	4.7 ± 1.6^{a}	109 ± 65.19^{a}	117 ± 67.09^{a}	50.9 ± 14.79^{a}	3.5 ± 1.7^{a}
BR	29.2 ± 3.40^{a}	8 ± 0.40^{a}	560.7 ± 211.8^{a}	278.7 ± 106.03^{a}	0.83 ± 0.55^{a}	143 ± 23.63^{a}	131.7 ± 19.29^{a}	59.4 ± 18.03^{a}	1.41 ± 0.3^{a}
SB	29.2 ± 1.40^{a}	8.4 ± 0.2^{b}	407.7 ± 81.1^{a}	203.7 ± 40.5^{a}	1.2 ± 0.32^{a}	124.3 ± 15.04^{a}	149.3 ± 52.0^{a}	55.1 ± 11.82^{a}	1.28 ± 0.2^{b}
SD	29.3 ± 1.25^{a}	8.9 ± 0.06^{b}	659.2 ± 490.3^{a}	329.7 ± 245.04^{a}	1.8 ± 0.32^{a}	148.7 ± 57.49^{a}	128.7 ± 54.63^{a}	211.8 ± 237.97^{a}	3.99 ± 1.9^{a}
MD	30.9 ± 1.40^{a}	8.6 ± 0.25^{a}	773.1 ± 572.1^{a}	384.7 ± 285.03^{a}	2.5 ± 0.98^{a}	165.3 ± 49.17^{a}	135.7 ± 13.65^{a}	350.7 ± 262.19^{a}	3.22 ± 2.4^{a}
TA	28.4 ± 0.58^{a}	8.6 ± 0.17^{b}	887 ± 223.1^{a}	550 ± 79.9^{a}	0.6 ± 0.21^{b}	167.7 ± 27.61^{a}	124.7 ± 9.45^{a}	62.6 ± 9.23^{a}	5.22 ± 0.4^{a}
TB	30.6 ± 0.13^{b}	8.5 ± 0.31^{a}	746 ± 70.2^{a}	373.3 ± 35.2^{a}	4.9 ± 4.92^{a}	146.7 ± 300.62^{a}	153.3 ± 30.35^{a}	64 ± 7.55^{a}	5.3 ± 1.4^{a}
KS	29.6 ± 0.43^{a}	8.5 ± 0.51^{a}	693.3 ± 64.9^{a}	346.7 ± 32.7^{a}	1.2 ± 0.25^{a}	154.3 ± 10.69^{a}	124.3 ± 10.02^{a}	67.3 ± 5.77^{a}	2.51 ± 0.8^{a}
SA	31.8 ± 0.36^{a}	8.1 ± 0.31^{a}	503 ± 122.5^{a}	351.7 ± 61.2^{a}	1 ± 0.32^{a}	148 ± 10.58^{a}	143.3 ± 33.26^{a}	63.7 ± 7.37^{a}	3.91 ± 0.6^{a}

Means with the same letter are not significantly different from each other (a = $P > 0.05$; b = $P < 0.05$).

Table 3.
Mean scores of selected physicochemical parameters of water during dry season.

Site	Taste	Odor	Temp°C	pH	EC	TDS	TURB	Alkalinity	Hardness	Salinity	DO (mg/1)
KL1	Unob	Unob	27	6.2	11.91	8	1	115	86	33	2.6
KL2	Unob	Unob	25.1	5.3	456	231	4.1	142	112	46	5.9
KL3	Unob	Unob	28.3	6.7	88.6	48	3.2	31	22	32.4	6.2
BR1	Unob	Unob	22.8	5.2	769	388	0.1	146	114	64	3.4
BR2	Unob	Unob	21.2	6.5	578	289	0.6	131	118	34	3.81
BR3	Unob	Unob	23.1	6.1	352	176	0.8	117	79	39.1	4.6
SB1	Unob	Unob	26.3	6.4	329	169	0.5	102	64	33	4.33
SB2	OBJ	Unob	22.7	6.3	488	246	1.1	89	120	36.3	3.7
SB3	Unob	Unob	26.2	6.5	431	218	0.8	121	126	47	4.3
SD1	Unob	Unob	25.6	6.6	96.8	52	1.2	56	36	28.1	6.4
SD2	OBJ	Unob	24	6.5	934	469	1.1	146	97	329	6.45
SD3	Unob	Unob	25.2	6.7	971	488	0.8	132	122	98	4. 53
MD1	Unob	Unob	22.4	6.5	146.8	76	1.4	87	78	32	4.81
MD2	OBJ	Unob	26.3	6.5	921	459	1.2	141	94	384	3.42
MD3	OBJ	Unob	24.1	6.3	1273	636	1.1	183	106	476	7.4
TA1	Unob	Unob	26.1	6.4	964	487	0.2	121	91	38.5	6.5
TA2	OBJ	Unob	28.2	6.5	1072	538	0.3	145	89	62	7.3
TA3	Unob	Unob	26.4	6.5	641	641	0.4	115	79	48	7.12
TB1	Unob	Unob	27.1	6.7	834	419	7.5	131	96	36	6.51
TB2	OBJ	Unob	26.5	6.5	711	361	1.2	121	123	43	8.45
TB3	OBJ	Unob	28.1	6.4	713	359	0.7	87	93	39	6.5
KS1	OBJ	OBJ	26.5	6.8	689	348	0.9	126	106	48	4.35
KS2	Unob	Unob	24.8	6.5	768	389	0.6	118	116	37	3.4
KS3	Unob	Unob	27.6	5.8	642	321	0.4	112	88	39	5.6
SA1	Unob	Unob	27.5	5.5	469	239	0.5	121	102	32	5.2
SA2	Unob	Unob	26.1	6.5	648	328	1.1	119	83	36	6.5
SA3	Unob	Unob	28.5	6.1	412	207	0.5	127	132	41	5.7

Table 4.
Physicochemical properties of underground water in Jalingo Metropolis during rainy season.

All sites in the study area during the dry season were within the WHO permissible limit of 6.5–8.5 except sites SB_1 (8.6) SD_1 (8.8) SD_2 (8.9) SD_3 (8.9), MD, (8.6) KS_1 (8.9) band KS_2 (8.6) these high values show alkalinity and it could be due to high concentration of basic cations e.g. magnesium, calcium, potassium etc. in the sampled sites.

All sites in the rainy season were within the WHO permissible limit of 6.5–8.5.

3.3 Electrical conductivity

Electrical conductivity is the ability of any medium, water in this case, to carry an electric current. The presence of dissolved solids such as calcium, chloride, and magnesium in water samples carries the electric current through water. Conductivity values obtained in this study showed that all the sample sites contained

Site	Temp°C	pH	EC	TDS	TURB	Alkalinity	Hardness	Salinity	DO (mg/1)
KL	26.8 ± 1.61 [a]	6.07 ± 0.71[a]	185.50 ± 237.37[a]	95.67 ± 118.89[a]	2.77 ± 1.60[a]	96 ± 57.89[a]	73.33 ± 46.32[a]	37.13 ± 7.68[a]	4.9 ± 1.99[a]
BR	22.35 ± 1.02[a]	5.93 ± 0.67[a]	566.33 ± 208.74[a]	284.33 ± 106.10[a]	0.5 ± 0.36[a]	131.33 ± 14.50[a]	103.67 ± 21.46[a]	45.7 ± 16.05[a]	3.94 ± 0.61[a]
SB	25.07 ± 2.05[a]	6.4 ± 0.1[b]	416 ± 80.55[a]	211 ± 38.97[a]	0.8 ± 0.3[a]	104 ± 16.09[a]	103.33 ± 34.20[a]	38.77 ± 7.32[a]	4.11 ± 0.36[a]
SD	24.93 ± 0.83[a]	6.6 ± 0.1[b]	667.27 ± 494.33[a]	336.33 ± 246.42[a]	1.03 ± 0.21[b]	111.33 ± 48.42[a]	85 ± 44.24[a]	151.7 ± 157.47[a]	6.43 ± 0.04[a]
MD	24.27 ± 1.96[a]	6.43 ± 0.12[b]	780.27 ± 576.14[a]	390.33 ± 286.24[a]	1.23 ± 0.15[b]	137 ± 48.12[a]	92.67 ± 14.05[a]	297.33 ± 234.35[a]	5.21 ± 2.02[a]
TA	26.9 ± 1.14[a]	6.47 ± 0.06[b]	892.33 ± 224.26[a]	555.33 ± 78.45[a]	0.3 ± 0.1[b]	127 ± 15.87[a]	86.33 ± 6.43[a]	49.5 ± 11.82[a]	6.97 ± 0.42[a]
TB	27.23 ± 0.81[a]	6.53 ± 0.15[b]	752.67 ± 70.44[a]	379.67 ± 34.08[a]	3.13 ± 3.79[a]	113 ± 23.07[a]	104 ± 16.52[a]	39.33 ± 3.51[a]	7.15 ± 1.12[a]
KS	26.3 ± 1.41[a]	6.37 ± 0.51[a]	699.67 ± 63.67[a]	352.67 ± 34.24[a]	0.63 ± 0.25[b]	118.67 ± 7.02[a]	103.33 ± 14.19[a]	103.33 ± 5.86[a]	4.45 ± 1.10[a]
SA	27.37 ± 1.21[a]	6.03 ± 0.50[a]	509.67 ± 123.14[a]	258 ± 62.70[a]	0.7 ± 0.35[a]	122.33 ± 4.16[a]	105.67 ± 24.70[a]	36.33 ± 4.51[a]	5.8 ± 0.66[a]

Means with the same letter are not significantly different from each other ($a = P > 0.05$; $b = P < 0.05$).

Table 5.
Mean scores of physicochemical parameters of water during rainy season.

appreciable amount of dissolved ions (10.00–958.00 S/m). Sampling sites TA2 and MD3 contained 1,066 and 1,261 ohm/cm. The EC range obtained in this study is higher than the range of 62.03 to 70.11ohm/cm reported by Ikhuoriah and Oronsaye [19] in assessing the physicochemical characteristics and some heavy metals of Ossiomo River, Ologbo – a tributary of Benin River, Southern Nigeria. This high EC values observed in some sites may be as a result of the chemicals present in ionic form in the dump sites that sink into the ground. The EC of all sites fall within the WHO recommended limit of 500–1,500 ohm/cm.

3.4 Total dissolved solids (TDS)

These are the inorganic matters and small amounts of organic matter, which are present as solution in water. The study found TDS values from the drinking water samples are all within the maximum limit of 1000 mg/L. The highest TDS values of 637 mg/L and the lowest TDS values of 6 mg/L correspond to samples from KL1 and TA3, respectively. The highest TDS values were recorded from site MD3 (630) and TA3 (637). This is higher than the range of 37.67 mg/L to 476.67 mg/L reported by Moses and Ishaku [21] in the evaluation of physicochemical properties of well water qualities in selected villages in Zing Local Government Area of Taraba State, Nigeria. The standard or allowable value of the TDS set by WHO is 1000 mg/L. Therefore, the TDS values obtained in this study are within the WHO permissible standards.

3.5 Turbidity

Turbidity is the cloudiness of water caused by a variety of particles and is a key parameter in drinking water analysis. It is also related to the content of diseases causing organisms in water, which may come from surface runoff. The turbidity values obtained from the study sites range from 0.4–10.6. The standard recommended maximum turbidity limit, set by WHO for drinking water is 5 nephelometric turbidity units (NTU). The lowest turbidity values of 0.4 NTU and highest value of 10.6 NTU were found for samples from site TA1 and TB1, respectively (**Table 4**). The mean turbidity for sample locations revealed that KL and TB have values of 4.7 ± 1.6 and 4.9 ± 4.2. This may be as a result of the open nature of the water making it easy for impurities that could block light reception to flow into it. All sites in the study area except for sites KL2 and TB1 were below the WHO permissible levels of 5NTU.

3.6 Alkalinity

Alkalinity is the measure of the acid – neutralizing capacity of water. In most natural waters, it is due to the presence of carbonate ($CO3^{-}$), carbonate (HCO-) and hydroxyl anions. The alkalinity levels recorded in this study range from 39 mg/l – 204 mg/l as shown in **Table 2** with a mean range of 109 ± 65.19–167.7 ± 27.61 as shown in **Table 3** during the dry season. The value of alkalinity obtained during the rainy season range from 31 mg/l –183 mg/l as shown in **Table 4** with a mean range of 96 ± 57.89–131.33 ± 14.50 shown in **Table 5**. These are higher than the range of 4.4 ± 0.38 mg/c – 17.8 ± 0.25 mg reported by Dimowo [22] in an assessment of some physicochemical Parameters of River Ogun (Abeokuta, Ogun State, South Western Nigeria).

This study revealed that all sites were above the WHO permissible levels of 1.0 – 100 mg.c except sites KL_3 (39) and SD, (84) during the dry season and sites KL_3 (31), MD, (87), SD, (56) SB_2 (89) and TB_3 (87). The high alkalinity levels could be due to high concentration of basic cations such as calcium (Ca), magnesium (Mg), Sodium (Na) and Potassium (K).

3.7 Hardness

The hardness recorded in this study ranged from (22 to 132) mg/l as shown in **Table 4** with a mean value range of (73.33 to 105,67)mg/l as shown in **Table 5** during the rainy season and the value range of (48 to 187)mg/l as shown in **Table 2** with a mean value of (117 to 153)mg/l as shown in **Table 3** during the dry season. In this study only three sampled sites $KL_{3,}$(48 mg/l), SD_1(67 mg/l) and SB_1(90 mg/l) recorded hardness within the WHO permissible limit of 100 mg/l. The remaining sampled sites are above the WHO (100 mg/l) permissible limit while during the rainy season, fifteen of the sampled sites water hardness ranged from(22 to 97)mg/l as shown in **Table 4** are within the WHO permissible limit of 100 mg/l and twelve of the sampled sites water hardness ranged from(102 to 132)mg/l are above the WHO permissible limit of 100 mg/l These high values could be due to the high concentrations of dissolved salts such as $MgSO_4$, $CaSO_4$, $CaCO_4$ found in rocks and soils. Most commonly associated with the ability of water to precipitate soap. As hardness increases, more soap is needed to achieve the same level of cleaning due to the interactions of the hardness ions with the soap. Chemically, hardness is the sum of polyvalent caution concentrations dissolved in water. In fresh waters, the principal hardness-causing ions are Calcium and Magnesium; Strontium, Iron, Barium and Manganese ions also contribute. The hardness recorded in this study range from 48 mg/l – 187 mg/l (**Table 4**), and a mean value of 117 $\pm$ 67.09 mg/l – 153.3 $\pm$ 30.35 mg/l. This is above the mean hardness range of 45.5 $\pm$ 4.79 mg/l – 105.0 $\pm$ 46.74 mg/l reported by Dimowo [22] in an assessment of some physico-chemical parameters of River Ogun (Abeokuta, Ogun State, Southwestern Nigeria) in Comparison with National and International Standards. These values exceed the maximum permissible WHO standards of 100 mg/l. this could be due to high concentrations of dissolved salt such as $MgSO_4$, $CaCO_3$, found in rocks and soils.

3.8 Salinity

The salinity of the samples collected from the sites in the study area range from 42 mg/l – 508 mg/l. The high values recorded were from sample sites SD2 (483 mg/l), MD2 (496 mg/l) and MD3 (508 mg/l). The salinity values for all sample sites are higher than the value range of 1.36 $\pm$ 0.13 mg/l to 5.27 $\pm$ 0.21 mg/l, with a mean value of 2.55 $\pm$ 0.109 mg/l reported by Bolarinwa, Fasakin and Fagbenro [23] in the analysis of the physicochemical parameters of coastal waters of Ondo State, Nigeria. The results obtained in this study are within WHO's permissible range of $\leq$600 mg/L.

Salinity: The salinity of water samplwes collected from the sites in the study area ranged from (28.1 476nmg/l) as shown in **Table 4** with a mean value of 36.33 $\pm$ 4.51–297.33 $\pm$ 234.35 during the rainy season and the value range of (33.4 – 508 mg/l) as shown in table with a mean value of 50.9 $\pm$ 14.79–350.7 $\pm$ 262.19 as shown in **Table 5** the salinity levels obtained in this study revealed that both in the dry and rainy season the results were within the WHO permissible limit of $\leq$600 mg/l.

3.9 Iron (Fe^{2+})

The Fe^{2+} concentrations recorded for all sites range between 0.01–0.08 mg/l. The mean concentration of Fe^{2+} range from 0.007–0.047 mg/l in all sample sites in the study area as shown in **Table 6**. These values are lower than the range of

Site	Fe^{2+}	Mn^{2+}	F^{-}	$NO3^{-}$	$NO2^{-}$	Cl^{-}	$SO4^{2-}$
KL	0.02 ± 0.01^{b}	0.023 ± 0.01^{a}	0.13 ± 0.15^{b}	0.56 ± 0.90^{a}	0.01 ± 0.01^{a}	27.17 ± 13.05^{a}	26.33 ± 7.37^{a}
BR	0.02 ± 0.01^{b}	0.07 ± 0.09^{b}	0.08 ± 0.07^{b}	0.98 ± 0.83^{a}	0.03 ± 0.01^{b}	99.67 ± 23.86^{a}	47 ± 10.54^{a}
SB	0.02 ± 0.05^{a}	0.09 ± 0.09^{b}	0.12 ± 0.11^{b}	0.09 ± 0.04^{b}	0.02 ± 0.01^{a}	121.47 ± 105.94^{a}	54.67 ± 33.55^{a}
SD	0.04 ± 0.02^{b}	0.03 ± 0.02^{b}	0.04 ± 0.02^{b}	0.4 ± 0.41^{a}	0.03 ± 0.01^{a}	112 ± 87.07^{a}	51.33 ± 37.63^{a}
MD	0.06 ± 0.01^{b}	0.02 ± 0.01^{a}	0.28 ± 0.38^{a}	13.56 ± 21.88^{a}	0.08 ± 0.04^{b}	185.43 ± 140.84^{a}	53 ± 22.65^{a}
TA	0.02 ± 0.01^{b}	0.08 ± 0.09^{b}	0.04 ± 0.01^{b}	0.86 ± 1.49^{a}	0.06 ± 0.03^{b}	110.33 ± 237.29^{a}	46.33 ± 13.05^{a}
TB	0.05 ± 0.03^{b}	0.03 ± 0.02^{b}	0.5 ± 0.42^{a}	7.8 ± 6.52^{a}	0.09 ± 0.11^{b}	221.67 ± 135.57^{a}	59.67 ± 7.57^{a}
KS	0.06 ± 0.04^{b}	0.02 ± 0.01^{a}	0.31 ± 0.28^{a}	1.54 ± 0.53^{a}	0.02 ± 0.01^{a}	91.67 ± 88.635^{a}	61.67 ± 21.39^{a}
SA	0.02 ± 0.02^{b}	0.02 ± 0.02^{b}	0.24 ± 0.24^{a}	0.48 ± 0.39^{a}	0.02 ± 0.01^{a}	43 ± 2^{a}	49.67 ± 9.29^{a}

Means with the same letter are not significantly different from each other ($a = P > 0.05$; $b = P < 0.05$).

Table 6.
Mean values of ions in water in Jalingo during rainy season.

0.15 mg/l and 3.26 mg/L reported by Popoola, Yusuf and Aderibigbe [24] in assessment of natural groundwater physico-chemical properties in major industrial and residential locations of Lagos metropolis. The WHO recommended limit for in drinking water is 0.3 mg/l. Therefore, the Fe concentrations recorded in the sampled water are just traces and not harmful to life.

3.10 Manganese (Mn^{2+})

The concentrations of Mn^{2+} from the study range from 0.01 mg/l – 0.17 mg/l in the sampled water sites as shown in **Table 7**. All the values obtained from the sample sites except for sites BR1 (0.12 mg/l), SB2 (0.15 mg/l) and TA3 (0.17 mg/l) are within the WHO [25] permissible levels of 0.1 mg/l to 0.5 mg/l. The mean concentration of Mn^{2+} ranged from(0.003–0.063) mg/l in all the sample sites as reported in **Table 6**. The

Site	Fe^{2+}	Mn^{2+}	F^-	$NO3^-$	$NO2^-$	Cl^-	$SO4^{2-}$
KL1	0.01	0.03	0.03	1.6	0.01	13	18
KL2	0.01	0.02	0.05	0.05	0.02	29.8	29
KL3	0.03	0.02	0.3	0.02	0.01	38.7	32
BR1	0.01	0.17	0.02	1.3	0.02	83	57
BR2	0.03	0.02	0.05	1.6	0.02	127	36
BR3	0.02	0.02	0.16	0.03	0.04	89	48
SB1	0.02	0.02	0.24	0.05	0.02	52	16
SB2	0.03	0.19	0.08	0.09	0.03	243.4	76
SB3	0.02	0.05	0.04	0.13	0.01	69	72
SD1	0.01	0.02	0.03	0.04	0.03	17	9
SD2	0.09	0.05	0.02	0.84	0.03	188	81
SD3	0.01	0.02	0.06	0.32	0.02	131	64
MD1	0.08	0.03	0.71	1.81	0.09	38.3	29
MD2	0.04	0.02	0.03	0.06	0.03	199	74
MD3	0.06	0.01	0.09	38.8	0.11	319	56
TA1	0.01	0.03	0.05	0.06	0.03	38	61
TA2	0.02	0.02	0.03	0.98	0.09	252	42
TA3	0.03	0.19	0.042	1.53	0.05	41	36
TB1	0.02	0.05	0.48	1.9	0.13	69	55
TB2	0.08	0.02	0.93	14.8	0.11	328	63
TB3	0.04	0.01	0.09	6.7	0.03	268	61
KS1	0.09	0.02	0.6	1.8	0.02	194	75
KS2	0.06	0.03	0.29	0.93	0.02	42	37
KS3	0.02	0.02	0.04	1.9	0.01	39	73
SA1	0.04	0.01	0.02	0.04	0.02	43	54
SA2	0.02	0.02	0.5	0.8	0.01	45	39
SA3	0.01	0.04	0.2	0.6	0.02	41	56

Table 7.
Ions from water in Jalingo Metropolis during rainy season.

highest value of 0.17 mg/l was recorded for site TA3, which is higher than the value of (0.057–0.010)mg/l reported by Ayeki, Asikhia and Ojeh [26] in the study of seasonal and spatial variation in physicochemical & biological quality of rainwater in Benin City, Edo State. All the mean values obtained from the locations were within the WHO [27] permissible levels of (0.1 to 0.5)mg/l.

Manganese (Mn^{2+}): The concentration of manganese from the study range from (0.00–0.17)mg/l in the sampled water as shown in **Table 2** with mean value of 0.01 ± 0.02 – o.6 $\pm$ 0.09 as shown in **Table 8** during the dry season and value range of (0.01–0.19)mg/l as shown in **Table 7** with the mean value range of 0.02 ± 0.02–- 0.09 ± 0.09 mg/l as show in **Table 6** during the rainy season were all within the WHO permissible levels of 0.1 mg/l to 0.5 mg/l. The highest value of 0/17 mg/l recorded for site TA_3 and 0.19 mg/l in dry season and rainy season respectively are higher than the value of (0.10–0.57)mg/l.

3.11 Fluoride(F^-)

The fluoride concentration recorded in this study range from(0.01–0.82)mg/l. The high levels were recorded in sample sites TB1 (0.4 mg/l), KS1 (0.4 mg/l) and TB2 (0.82 mg/l). The permissible range of fluoride by WHO is(0.5 to 1.5)mg/l. The fluoride concentrations found in the study are within the permissible WHO standards. Therefore, the fluoride found in the drinking water in the study area is not detrimental to life.

3.12 Nitrate (NO_3^-)

The nitrate (NO_3^-) levels obtained in the study range from 0.01–38.5 mg/l for the sample sites. The high values of (6.2 mg/l),(14.4mg/l) and (38.5 mg/l were observed at site TB3, TB2 and MD3, respectively. This is higher than the range of 0.33 and 2.37 mg/L reported by Popoola, Yusuf and Aderibigbe [24] in assessment of natural groundwater physico-chemical properties in major industrial and residential locations of Lagos metropolis. All the nitrate values obtained in the study are lower than the WHO [27] recommended guideline value of 50 mg/l. This implies that the nitrate concentrations obtained in the study are within the permissible standards and might not pose any health risk.

Nitrate (NO_3^-) The nitrate levels obtained in this study range from 0.01–38.5 mg/l as shown in **Table 9** with mean value range of 0.06 ± 0.03–7.4 ± 6.48 as shown in **Table 8** during the dry season and the value range of 0.02–38.8 mg/l as shown in **Table 7** with mean value of range of 0.09 ± 0.04–13.56 ± 21.88 as shown in **Table 6** during the rainy season.

The high values of TB_3 (6.7), TB_{2+} (14.8) and MD_3 (38.8) during thee rainy season and TB_3 (6.2 mg/l) TB_2 (14.4 mg/l) and MD_3 were higher than the range of 0.33 and 2.37 mg/l reported by Popoola, Yusuf and aderigbe [24] in assessment of natural groundwater physicochemical properties in major industries and residential locations of Lagos metropolis.

All the Nitrate values obtained during the dry and rainy season are lower than the WHO [27] recommended guidelines value of 50 mg/l which implies that the nitrate concentrations obtained in the study were within the WHO standard and might not cause any health risk (**Tables 10** and **11**).

3.13 Nitrite (NO_2^-)

The nitrite (NO_2^-) levels obtained in this study range from 0.01 mg/l – 0.09 mg/l for all the sampled sites. The highest value of 0.09 mg/l was observed at site TB2 and MD3. This is lower than the value of 0.27 ± 0.005 mg/l reported by Bolarinwa,

Site	Fe^{2+}	Mn^{2+}	F^-	$NO3^-$	$NO2^-$	Cl^-	$SO4^{2-}$
KL	0.007 ± 0.01^a	0.013 ± 0.02^b	0.047 ± 0.05^b	0.48 ± 0.79^a	0 ± 0^b	24.17 ± 12.69^a	33 ± 9.8^a
BR	0.01 ± 0.01^b	0.043 ± 0.07^b	0.05 ± 0.04^b	0.87 ± 0.74^a	0.017 ± 0.01^a	94.67 ± 22.3^a	52 ± 9.5^a
SB	0.013 ± 0.01^a	0.053 ± 0.08^b	0.097 ± 0.09^b	0.06 ± 0.03^b	0.01 ± 0.01^a	117.13 ± 106.2^a	63 ± 36.4^a
SD	0.027 ± 0.04^b	0.01 ± 0.02^b	0.02 ± 0.02^b	0.28 ± 0.31^a	0.013 ± 0.01^a	107.33 ± 86.9^a	64 ± 43.4^a
MD	0.047 ± 0.04^b	0.01 ± 0.01^b	0.023 ± 0.34^a	13.43 ± 21.7^a	0.057 ± 0.04^b	179.93 ± 137.9^a	69.33 ± 22.0^a
TA	0.013 ± 0.02^b	0.063 ± 0.09^b	0.028 ± 0.01^a	0.82 ± 0.72^a	0.043 ± 0.04^b	104 ± 118.7^a	56.67 ± 14.4^a
TB	0.03 ± 0.03^b	0.017 ± 0.02^b	0.433 ± 0.37^a	7.4 ± 6.48^a	0.063 ± 0.04^b	214.33 ± 135.2^a	79.33 ± 7.6^a
KS	0.037 ± 0.04^b	0.003 ± 0.03^a	0.21 ± 0.19^b	1.33 ± 0.46^a	0.007 ± 0.01^a	86 ± 82.3^a	74.67 ± 25.9^a
SA	0.01 ± 0.01^b	0.007 ± 0.01^b	0.13 ± 0.15^b	0.41 ± 0.39^a	0.003 ± 0.01^a	62.67 ± 2.08^a	62.67 ± 4.5^a

Means with the same letter are not significantly different from each other ($a = P > 0.05$; $b = P < 0.05$).

Table 8.
Mean values of selected ions in water in Jalingo during dry season.

Site	Fe^{2+}	Mn^{2+}	F^-	$NO3^-$	$NO2^-$	Cl^-	$SO4^{2-}$
KL1	0.01	0.01	0.01	1.4	0	10	22
KL2	0.01	0.03	0.03	0.04	0	28	36
KL3	0	0	0.1	0.01	0	34.5	41
BR1	0.02	0.12	0.03	1.2	0.01	78	61
BR2	0.01	0.01	0.02	1.4	0.03	120	42
BR3	0	0	0.1	0.02	0.01	86	53
SB1	0.01	0.01	0.2	0.03	0	48	21
SB2	0.02	0.15	0.06	0.07	0.01	239.4	82
SB3	0.01	0	0.03	0.09	0.02	64	86
SD1	0	0	0.01	0.02	0.01	12	14
SD2	0.07	0.03	0.01	0.62	0.02	182	92
SD3	0.01	0	0.04	0.21	0.01	128	86
MD1	0.06	0.01	0.62	1.76	0.07	36.8	48
MD2	0	0	0.01	0.04	0.01	191	92
MD3	0.08	0.02	0.07	38.5	0.09	312	68
TA1	0.01	0.01	0.03	0.04	0.01	32	73
TA2	0.03	0.01	0.02	0.96	0.08	241	51
TA3	0	0.17	0.034	1.46	0.04	39	46
TB1	0.01	0.04	0.4	1.6	0.08	62	71
TB2	0.06	0.01	0.82	14.4	0.09	320	86
TB3	0.02	0	0.08	6.2	0.02	261	81
KS1	0.07	0	0.4	1.6	0.01	181	93
KS2	0.04	0.01	0.21	0.8	0.01	39	45
KS3	0	0	0.02	1.6	0	38	86
SA1	0.02	0	0	0.02	0	41	63
SA2	0.01	0	0.3	0.4	0	42	58
SA3	0	0.02	0.1	0.8	0.01	38	67

Table 9.
Ions from water in Jalingo Metropolis during dry season.

Fasakin and Fagbenro [23] in the analysis of the physicochemical parameters of coastal waters of Ondo State, Nigeria. All the nitrite values obtained in the study are lower than the WHO [27] recommended guideline value of 50 mg/l. This implies that the nitrite concentrations obtained in the study are only traces and might not cause any health risk to human life.

3.14 Chloride(Cl^-)

The chloride levels recorded in this study range from 10 mg/l – 320 mg/L.The high values were recorded in three sites; TB3 (261 mg/l), MD3 (312 mg/l) and TB2 (320 mg/l). The values for all sample sites are higher than the range of 0.23 mg/L – 9.0 mg/L and mean concentration of 1.81 mg/L reported by Moses and Ishaku [21] in the evaluation of physicochemical properties of well water qualities in selected villages in Zing Local Government Area of Taraba State, Nigeria. This high values of

Parameter	Well	Borehole	Pipewater	WHO	NSDW
Taste	Unob	Unob	OBJ	—	—
Odor	Unob	Unob	Unob	—	—
Temp °C	29.79 ± 0.01^b	29.63 ± 0.01^b	30.7 ± 0.02^b	40	
pH	8.37 ± 0.8^a	8.34 ± 0.12^a	8.7 ± 0.1^a	6.5–8.5	—
EC	289.5 ± 0.6^b	766 ± 0.02^b	708 ± 0.01^b	500–1500	1000
TDS	45 ± 0.2^a	383 ± 0.61^b	354 ± 2.36^b	500	500
TURB	4.8 ± 2.5^a	0.2 ± 0.19^a	2.4 ± 4.26^b	5NTU	
Alkalinity	39 ± 0.3^b	168 ± 0.5^b	$130 \pm 0.1.6^b$	100	100
Hardness	48 ± 0.13^b	147 ± 0.017^b	186 ± 0^b	100	150
Salinity	42.8 ± 1.2^b	78 ± 0.2^b	71 ± 0.02^b	≤600	

Means with the same letter are not significantly different (a = P > 0.05; b = P < 0.05).

Table 10.
Variation in selected physicochemical parameters as influenced by water sources.

Water Source	Well	Borehole	Pipe water	WHO	NSDW
Fe	0.18 ± 0.02^b	0.12 ± 0.01^b	0.06 ± 0.01^b		
Mn	0.11 ± 0.07^b	0.08 ± 0.01^b	0.01 ± 0.01^b	0.05	
F	0.17 ± 0.01^b	0.13 ± 0.01^b	0.82 ± 0.02^a	0.5–1.50	
NO3	2.81 ± 0.8^a	0.6 ± 0.12^a	14.4 ± 1.6^a	50	
NO2	0.16 ± 0.6^b	0.11 ± 0.02^b	0.09 ± 0.01^b	50	
Cl	265.71 ± 0.2^a	269.1 ± 0.61^a	320 ± 18.36^a		
SO4	55.1 ± 2.5^a	78.8 ± 0.19^a	86 ± 4.96^a	400	100
DO	2.97 ± 0.3^a	4.85 ± 0.43^a	$6.87 \pm 0.1.6^a$	≥6	—

Means with the same letter are not significantly different (a = P > 0.05; b = P < 0.05).

Table 11.
Variation in selected ions as influenced by water sources.

chloride found in the study area which is more than the WHO permissible limit of 250 mg/l could be due to the high concentration of chloride in the rocks, soils and spraying of agro-chemicals like DDT and atrazine.

3.15 Sulphate (SO_4^{2-})

The sulphate (SO_4) concentration of the samples collected from the sites in the study area range from 14 mg/l – 93 mg/l. The concentration is below the range of 3.10 mg/l to 66.10 mg/L reported by Moses and Ishaku [21] in the evaluation of physicochemical properties of well water qualities in selected villages in Zing Local Government Area of Taraba State, Nigeria. In contrast, the range found in the present study is higher the range of 13–63 mg/L reported by Popoola, Yusuf and Aderibigbe [24] in assessment of natural groundwater physico-chemical properties in major industrial and residential locations of Lagos metropolis. The permissible values by WHO and NSDWQ are 250 mg/L and 100 mg/L respectively. Therefore, the sulphate concentrations found in the water from the sampled sites are within the permissible limit of WHO standards for good quality water.

3.16 Dissolved oxygen (DO)

Dissolved Oxygen (DO) levels from the study revealed a range of 1.1 mg/l – 6.87 mg/l and 2.6–8.45, and mean range of 1.28 ± 0.2–5.3 ± 1.4 and 3.94 ± 0.61–7.15 ± 1.12 for dry and rainy seasons respectively. This is lower than the values of 6.51 ± 1.47 mg/l – 8.43 ± 0.21 mg/l reported by Bolarinwa, Fasakin and Fagbenro [23] in the analysis of the physicochemical parameters of coastal waters of Ondo State, Nigeria. The WHO permissible standard for DO is >6 mg/l. Most of the dissolved oxygen concentrations found in water from the sampled sites do not meet the WHO standard for quality drinking water. This could be as a result of the high concentrations of dissolved salts, suspended particles and hardness of the water that preclude the dissolution of oxygen in most of the water from the sampled sites.

The water samples for this study were collected from three main sources. These include well, borehole and pipewater. The taste of water samples from the different sources differ as the pipewater is objectionable as compared to well and borehole sources, which are unobjectionable. In terms of the odor, water from all the sources recorded no odor. The temperature difference across the water sources is not statistically significant. pH across the sources is not significantly different. There is significant difference in Total Dissolved Solids and Turbidity across the various water sources. All other variables are not significantly different across the all sources.

The selected metal ions in water from the three sources are presented in **Table 9**. There was significant difference recorded in fluoride (F^-) across the water sources. All other variables recoded no significant difference across the different water sources in the study area. This implies that, there is not marked difference between the composition and concentration of physicochemical parameters in water from all water sources.

3.17 Discussion

The water temperature from all sites ranges between 26.7°C – 33. The temperature values obtained from this study were within the WHO permissible limit of 40°C. This is in support of the work by Moerman, Blokker, Vreeburg, and Van Der Hoek [28] that water temperature within the range of 28 – 380c is consider safe for drinking.

pH measure the acidity or alkalinity of the water. The pH values from this range from (6.5–8.9) also reported by Ikhouriah and Oronsaye [19] in Ossiomoo river Ologbo - a tributary of Benin River, Southern Nigeria. All sites in the study area during the dry season were within the WHO permissible limit of 6.5–8.5 except sites SB_1 (8.6) SD_1 (8.8), SD_2 (8.9) SD_3 (8.9), MD, (8.6), KS_1 (8.9) band KS_2 (8.6) these high values indicates alkalinity, and it could be due to high concentration of basic cations, e.g. magnesium, calcium, potassium etc. in the sampled sites. Electricity conductivity values obtained in this study showed that all the sample sites contained an appreciable amount of dissolved ions (10.00–958.00 S/m). The EC of all sites falls within the WHO recommended limit of 500–1,500 ohm/cm.

Turbidity is the impurity status of the water caused by a variety of particles and is a key parameter in drinking water analysis. It is also related to the content of diseases causing organisms in water, which may come from surface runoff. The turbidity values obtained from the study sites range from 0.4–10.6. The standard recommended maximum turbidity limit set by WHO for drinking water is five nephelometric turbidity units (NTU). These results from the open nature of the water, making it easy for impurities that could block light reception to flow into it. All sites in the study area except for sites two sites were below the WHO permissible levels of 5NTU.

The alkalinity levels recorded in this study range from 39 mg/l – 204 mg/l with a mean range of 109 ± 65.19–167.7 ± 27.61.during dry season and 31 mg/l–183 mg/l, with a mean range of 96 ± 57.89–131.33 ± 14.50 during the wet season. These are higher than the range of 4.4 ± 0.38 mg/c – 17.8 ± 0.25 mg reported by Dimowo [22] in an assessment of some physicochemical Parameters of River Ogun (Abeokuta, Ogun State, South-Western Nigeria). All sites were above the WHO permissible levels of 1.0 – 100 mg.c except for sites few sites. The high alkalinity levels could be due to a high concentration of basic cations such as calcium (Ca), magnesium (Mg), Sodium (Na) and Potassium (K).

Salinity: The salinity of water samples collected from the sites in the study area ranged from (28.1 mg/l to 476 mg/l) as shown in **Table 4** with a mean value of 36.33 ± 4.51 mg/l – 297.33 ± 234.35 mg/l during the rainy season and the value range of (33.4 mg/l – 508 mg/l) as shown in **Table 2** with a mean value of 50.9 ± 14.79–350.7 ± 262.19 as shown in **Table 3** during the dry season. The salinity values for all sample sites are higher than the value range of 1.36 ± 0.13 mg/l to 5.27 ± 0.21 mg/l, with a mean value of 2.55 ± 0.109 mg/l reported by Bolarinwa, Fasakin and Fagbenro [23] in the analysis of the physicochemical parameters of coastal waters of Ondo State, Nigeria. The salinity levels obtained in this study revealed that both in the dry and rainy season, the results are within the WHO permissible limit of ≤600 mg/l.

The Fe^{2+} concentrations recorded for all sites range between 0.01 mg/l – 0.08 mg/l The values obtained in the study are lower than the range of 0.15 mg/l and 3.26 mg/L reported by Popoola, Yusuf and Aderibigbe [24] in the assessment of natural groundwater physicochemical properties in major industrial and residential locations of Lagos metropolis. The values of Fe recorded both in dry and rainy seasons are within the WHO permissible limit of 0.3 mg/l. Therefore, the Fe concentrations recorded in the sampled water are just traces and not harmful to life.

The concentration of manganese from the study range from (0.0–0.17)mg/l. The highest value of 0.17 mg/l and 0.19 mg/l recorded for site TA_3 in the dry season and rainy season, respectively, are higher than the value of (0.010–0.057)mg/l reported by Ayeki, Asikhia and Ojeh [26] in the study of seasonal and spatial Variation in Physico-chemical and biological quality of water in Benin City, Edo State, Nigeria. All the manganese values obtained from the study are within the WHO [27] permissible standard of (0.1 to 0.5)mg/l.

The Nitrate (NO_3^-) levels obtained in this study ranged from 0.01–38.5 mg/l during the wet and dry season. All the Nitrate values obtained during the dry and rainy seasons are lower than the WHO [27] recommended guidelines value of 50 mg/l, which implies that the nitrate concentrations obtained in the study are within the acceptable limit and might not cause any health risk.

The nitrite (NO_2^-) levels obtained in this study range from 0.01 mg/l – 0.09 mg/l with a mean value range of 0.01–0.063 during the dry season and the levels obtained during the rainy season ranged from 0.01 mg/l to 0.13 mg/l with a mean value range of 0.01 mg/l to 0.09 mg/l are all within the WHO [27] permissible nitrite limit of 50 mg/l. The highest value of 0.09 mg/l was observed at site TB2 and MD3. This is lower than the value of 0.27 ± 0.005 mg/l reported by Bolarinwa, Fasakin and Fagbenro [23] in the analysis of the physicochemical parameters of coastal waters of Ondo State, Nigeria. This implies that the nitrate concentrations obtained in the study are only traces and might not cause any health risk to human life.

The chloride levels recorded in this study range from 10 mg/l – 320 mg/L with a mean value range of 24.17 mg/l to 2114.33 mg/l during the dry season and the value range from 13 mg/l to 328 mg/l with a mean value range of 27.17 mg/l to 221.67 mg/l during the rainy season. The high values were recorded in three sites; TB3 (261 mg/l),

MD3 (312 mg/l) and TB2 (320 mg/l) during the dry season and TB3(268 mg/l), MD3 (319 mg/l) and TB2(328 mg/l) during the rainy season are higher than the WHO permissible limit of chloride (250 mg/l) but the rest of the sampled sites are within the permissible limit of the WHO. The high values of chloride found in the study area could be due to the high concentration of chloride in the rocks, soils and spraying of agrochemicals like DDT and atrazine.

The sulphate (SO_4) concentrations of the sampled water range from 14 mg/l – 93 mg/l with the mean range of 33 mg/l to 79.33 mg/l during the dry season and the levels during the rainy season ranged from 9 mg/l to 81 mg/l with a mean value range of 26.33 mg/l to 61.67 mg/l. The concentrations recorded in this study are below the range of 3.10 mg/l to 66.10 mg/L reported by Moses and Ishaku [21] in the evaluation of physicochemical properties of well water qualities in selected villages in the Zing Local Government Area of Taraba State, Nigeria. In contrast, the range found in the present study is higher than the range of 13–63 mg/L reported by Popoola, Yusuf and Aderibigbe [24] in the assessment of natural groundwater physicochemical properties in major industrial and residential locations of the Lagos metropolis. The permissible level of WHO for sulphate is 100 mg/l and the concentrations recorded in this study both in dry and rainy seasons are within the permissible limit of the WHO standard for good quality water.

Dissolved Oxygen (DO) levels from the study revealed a range of 1.1 mg/l – 6.87 mg/l and 2.6–8.45, and a mean range of 1.28 ± 0.2–5.3 ± 1.4 and 3.94 ± 0.61–7.15 ± 1.12 for dry and rainy seasons, respectively. This is lower than the values of 6.51 ± 1.47 mg/l – 8.43 ± 0.21 mg/l reported by Bolarinwa, Fasakin and Fagbenro [23] in the analysis of the physicochemical parameters of coastal waters of Ondo State, Nigeria. The WHO permissible standard for DO is >6 mg/l. Most of the dissolved oxygen concentrations found in water from the sampled sites do not meet the WHO standard for quality drinking water. This could be as a result of the high concentrations of dissolved salts, suspended particles and hardness of the water that preclude the dissolution of oxygen in most of the water from the sampled sites.

4. Conclusion

The study set out to undertake the physicochemical analysis of water quality from multiple sources in Jalingo Metropolis, Nigeria. In Nigeria today, the use of ground water has become an agent of development because the government is unable to meet the ever increasing demand of water supply.

Water from natural sources is never completely pure. Most of the earth's water sources get their water supplies through precipitation. During precipitation water passes over (runoff) and through the ground (infiltration), acquiring a wide range of dissolved or suspended impurities that intensely alters its usefulness. The physicochemical parameters identified include temperature, electrical conductivity, total dissolved solids, manganese (Mn), iron (Fe), nitrate (NO3), sulphate (SO4), salinity, fluoride (F) and nitrite (NO2), which were within the permissible standards of WHO. Other parameters examined included pH, turbidity, chloride, hardness, alkalinity and dissolved oxygen, which were either lower or higher values outside the permissible standards of WHO and NSDW.

The study sampled three sources; well, borehole and tap water. The findings revealed that the tastes of water samples from the different sources differ. Tap water was found to be objectionable as compared to well and borehole sources. All the water sources recorded no odor. Physicochemical parameters of Total Dissolved Solids, Turbidity, and fluoride (F^-) recorded significant difference across the various water sources. All other variables recorded no significant difference across the different water sources in the study area.

5. Recommendations

Based on the findings of this study, the following recommendations are made:

i. There should be frequent water and treatment within Jalingo metropolis because of the bad practices of the indiscriminate waste disposal, excessive use of agro-chemicals that could contaminate the water sources.

ii. Policy makers and non-governmental organizations should intensify efforts in providing water that is safe for human consumption as some of the physicochemical parameters of the water in the study area (Jalingo Metropolis) are outside the recommended limits of WHO.

iii. Health officials and regulatory bodies should ensure that boreholes are not indiscriminately sited especially close to dumpsites and domestic sewages.

iv. Government should enforce the regulations on the excessive use of chemical fertilizers and on the indiscriminate waste disposal.

Author details

Benjamin Ezekeil Bwadi[1*], Mohammed Bakoji Yusuf[1], Ibrahim Abdullahi[1], Clement Yakubu Giwa[2] and Grace Audu[1]

1 Department of Geography, Taraba State University, Jalingo, Nigeria

2 Department of Geography, College of Education Zing, Zing, Nigeria

*Address all correspondence to: bwadiben@gmail.com

References

[1] Yusuf, Y.O. and Shuaib, M.I. (2012). The Effect of Wastes Discharge on the Quality of Samaru Stream, Zaria, Nigeria, *Ecological Water Quality - Water Treatment & Reuse,* Dr Voudouris *(Ed.),* Effect of wastes discharge on the quality of Samaru stream in Zaria, Nigeria.

[2] Jidauna, G.G., Dabi, D.D., Saidu, B.J., Ndabula, C. and Abaje, I.B. (2014). Chemical Water Quality Assessment in Selected Location in Jos, Plateau State, Nigeria. Research Journal of Environmental and Earth Sciences, **6** (5), 284-291.

[3] Dabi, D.D. and Jidauna, G.G. (2010). Climate change and local perception in selected settlement in the Sudano-Sahelian region of Nigeria. Journal of Environmental Sciences and Resources Management, **2**, 1-12.

[4] Odoh, R. and Jidauna, G.G. (2013). The Spatial Effect of Rusty Roof on Water Quality in Otukpo Local Government Area of Benue State, Nigeria *International Journal of Marine, Atmosphere, & Earth Science,* **1**(1), 27-37.

[5] Gimba, P.B. (2011). Assessment of quality of drinking water in Bosso town, Niger state. Master Thesis, Ahmadu Bello University, Zaria, Nigeria.

[6] Park's Textbook of Preventive and Social Medicine, 18th Edition 2005.519-541. M/s Banarsidas Bhanot publishers 1167, Prem Nagar, Jabalpur, India

[7] WHO and UNICEF JMP (2006). Meeting the MDG Drinking Water and Sanitation Target. The Urban and RuralChallenge of the Decade.

[8] Diersong, N. (2009). Water Quality; Frequently Asked Questions. Florida Brooks National Marine Sanctuary, Key West. pp. 1-2.

[9] United State Environmental Protection Agency (USEPA). (2005). Protecting Water Quality from Agricultural run-off, pp 1-5.

[10] Standard Organization of Nigeria, SON (2007). *Nigerian Standard for drinking water quality*. Lagos, Nigeria: Standard Organization of Nigeria.

[11] Reynolds, K.A., Mena, K.D. and Gerba, C.P. (2008). *Water Quality Analysis*. The University of Arizona, Mel and Enid Zuckerman College of Public Health, 1295 N. Martin Ave., Tucson, AZ 85724, USA.

[12] Organization, W. H., & UNICEF. (2008). Progress on drinking water and sanitation: special focus on sanitation *Progress on drinking water and sanitation: special focus on sanitation* (pp. 58-58).

[13] Ndabula, C. and Jidauna, G.G. (2010). Domestic water use in Selected Settlements in the Sudano-Sahelian Region of Nigeria. International Journal of Water and Soil Resources Research, **1** (1–3), 1-11.

[14] Tebbutt, T. H. Y. (1990). *Basic water and waste water treatment*. London: Bultrtwith.

[15] Oruonye, E. (2011). An assessment of the level of awareness of the effects of climate change among students of tertiary institutions in Jalingo Metropolis, Taraba State Nigeria. *Journal of Geography and Regional Planning, 4*(9), 513-517.

[16] Organization, W. H. (2021). Working for health: a review of the relevance and effectiveness of the five-year action plan for health employment and inclusive economic growth (2017-2021) and ILO-OECD-WHO Working for Health programme.

[17] Commission, N. P. (2021). National housing Demarcation Survey: NPC.

[18] Adejuwon J. O., and Adelakun M. A., 2012, Physiochemical and bacteriological analysis of surface water in Ewekoro Local Government Area of Ogun State, Nigeria: Case study of Lala, Yobo and Agodo Rivers. International Journal of Water Resources and Environmental Engineering, **4**(3), 66-72.

[19] Ikhuoriah, S.O. and Oronsaye, C.G. (2016). Assessment of Physicochemical Characteristics and some Heavy Metals of Ossiomo River, Ologbo – A Tributary of Benin River, Southern Nigeria. *J. Appl. Sci. Environ. Manage.*, **20**(2) 472 – 481.

[20] Chinedu S. N., Nwinyi O. C., Adetayo Y. O., and Eze V. N. (2011). Assessment of water quality in Canaanland, Ota, Southwest Nigeria. Agriculture and Biology Journal of North America, **2**(4), 577-583.

[21] Moses, A.N. and Ishaku, S. (2020). Evaluation of Physico-Chemical Properties of Well Water Qualities in Selected Villages in Zing Local Government Area of Taraba State, Nigeria. International Journal of Contemporary Research and Review, **11** (03), 20282–20288.

[22] Dimowo, B.O. (2013). Assessment of Some Physico-chemical Parameters of River Ogun (Abeokuta, Ogun State, Southwestern Nigeria) in Comparison With National and International Standards.International Journal of Aquaculture, **3**(15), 79–84.

[23] Bolarinwa, J.B., Fasakin, E.A. and Fagbenro, A.O. (2016). Physicochemical Analysis of the Coastal Waters of Ondo State, Nigeria. International Journal of Research in Agriculture and Forestry, **3** (11), 13-20.

[24] Popoola, L.T., Yusuf, A.S. and Aderibigbe, T.A. (2019). Assessment of natural groundwater physico-chemical properties in major industrial and residential locations of Lagos metropolis. Applied Water Science, **9** (191), 1–10.

[25] Organization, W. H. (2019). Global perspectives on assistive technology: proceedings of the GReAT Consultation 2019, World Health Organization, Geneva, Switzerland, 22–23 August 2019. Volume 1.

[26] Ayeki, J.A., Asikhia, M. and Ojeh, V. N (2018). Seasonal and Spatial Variation in PhysicoChemical & Biological Quality of Rainwater in Benin City, Edo State. *Global Journal of Human Social Science*, **18**(5)I, 25–33.

[27] Organization, W. H. (2012). *World malaria report: 2012*: World Health Organization.

[28] Moerman, A., Blokker, M., Vreeburg, J., & Van Der Hoek, J. (2014). Drinking water temperature modelling in domestic systems. Procedia Engineering, 89, 143-150.

Section 2

Factors of Water Quality Degradation

Chapter 5

Degradation and Improvement of Urban River Water Quality

Thomas D. Shahady

Abstract

The need to understand urban water quality has become a very important area of research and management in the aquatic sciences. Decades of urban development generating high rates of impervious surface, complex networks of stormwater control mechanisms and declining river water quality has created a demand for greater study. In this chapter, issues such as stream bank erosion, flooding, sediment pollution, bacteria and channelization are presented as drivers of the urban water environment. Methodologies and study designs to document these impacts are discussed. Ideas to improve the urban condition such as retrofitting previous development, infiltration of surface runoff, stream restoration, dredging and rehabilitation of lakes and compartmentalization of future development are explained and detailed as ways of integrating the natural landscape features into improvement of our urban centers. Finally, the incorporation of citizen science into adaptive policy is suggested as a solution to regulatory and esthetic/recreational need for improvement.

Keywords: Urbanization, Water Quality, Stormwater, Best Management Practices, Sediment Pollution

1. Introduction - how did we get here?

Urban environments including our buildings, roads, traffic, residences, industries, sewage treatment systems, population centers, parks and precipitation patterns all intertwine creating observed water quality in these communities. Water resources and the people living in and around them are intricately linked whether we understand this or not. The observed impact is now so pervasive scientists have suggested a new epoch beginning in the 1950s as the Anthropocene. Regardless of the label, this current age of human influence on the environment is very pervasive creating many stressors that plague water quality. Excessive sediment and erosion mixed with microplastics, pathogens, toxics and nutrients flow through our urban rivers daily. In order to improve this condition, we must first understand the origins.

Our rivers were not always degraded. Nature exists in a state of dynamic equilibrium with our surroundings. Rivers are no different. The river is in equilibrium with its drainage area. Historically, precipitation struck vegetation directly infiltrating into the ground as it was not covered by development. Entire watersheds were vegetated and these are the conditions rivers and streams equilibrated into. Precipitation flowed into the river but mainly through the surrounding soil first. The watershed stored precipitation in the ground, purified it as it slowly migrated toward the stream then released it through lateral discharge into our rivers. Any precipitation flowing over land only occurred during large storms and this input was periodic. Thus, the size and shape of streams and rivers reflected this pattern. The river system sometimes flooded but not as a catastrophic event. It flowed into the floodplain adjacent to its banks where essential nutrients and replenishing sediments sustained the river system. Rivers created a mosaic of serpentine meanders from mountain headwaters to the coast gently changing while moving sediment from one bank to another. It was a river in harmony with its surroundings.

But floodplains are flat and the alluvial soil desirable. Building projects need flat land for development and farmers need good soil for crops. These floodplains became premium resources early in our history before we learned to flatten plots of land for our buildings and fertilize soil for our crops. We developed floodplains around major rivers that soon became urban centers. Development moved up the water course fanning throughout the drainage basin. Rivers were our first highways, so commerce easily moved along these waterways and development prospered at major ports.

Precipitation contacting these built impervious surfaces instantaneously generated a new pattern of water flow. Water that previously traveled through groundwater now flowed as surface runoff. A river once in harmony with its surroundings and dependent on groundwater (a term called hyporheic flow) now became a conveyance for surface water discharge. In response, the river eroded to gain harmony with this *new* pattern of flow. Development increased, engineers designed more conveyances, pipes, levees and floodplain relevation to alleviate flooding and remove water quickly from our built surfaces. Now contaminated and not purified, surface water became the predominate input into our rivers. More built environments created ever increasing discharge resulting in dangerous flooding. To contain this problem, retaining walls were built. River erosion multiplied as construction continued throughout the watershed. As this intensified, encasement of the river into pipes and culverts became the standard of management. What was once a majestic river flowing through a beautiful landscape naturally purifying itself was transformed into an artificial conveyance filled with polluted water.

1.1 The paradox of development

Our social evolution further exacerbated this problem. The built environment is expensive and with the advent of indoor plumbing an extensive system of water movement became necessary. These pipelines were built under the construction then along the river banks impacting its natural flow. Large scale purification plants appeared at the river mouths. The extensive urban network required rapid and efficient movement of water away from surfaces so extensive systems of stormwater drainage began to appear. Environmental regulation multiplied requiring more and more infrastructure with development. The landscape became an intricate stormwater and sewage drainage network littered with culverts, sedimentation ponds, curb and gutter and isolated streams only noticeable over bridge crossings.

To pay for this, urban governments taxed the land that was developed. Zoning of land became commonplace and classification as improved or unimproved was designated. Governments believed building upon the land improved it because such lands generated greater tax revenue. But from an environmental perspective, these projects were not improvements but environmentally liabilities increasing in cost over time.

Thus the paradox. Often, urban governments have autonomy over local land decisions. As a result, property is taxed and local governments funded. Government revenue increases with land improvements creating the need to grow in order to meet increasing population demands. Local governments find the need to approve

Figure 1.
A river flowing through the urban landscape.

development to support its infrastructure and the cycle continues. It is difficult to suggest stopping this because you are asking government to stop funding the essential services you need such as police, fire protection, schools and roads.

As this development continued, groups of environmental researchers began to observe how rivers were responding. Leopold et al. [1] called it the urbanization cycle where heavy development caused rapid sedimentation followed by fundamental changes in river hydrology. In the same year, Walsh et al. [2] along with Meyer et al. [3] coined the term urban stream syndrome. The descriptions were alarming. With development, the stream channel begins to deepen and widen with the banks becoming unstable and eroding. Deepening of the river isolates it from the floodplain resembling a simple conveyance rather than an integrated stream course. Continued overdevelopment isolates remaining ecological services provided by trees, soil and wildlife. As waste and energy needs for the urban system intensifies, the cycle of erosion and isolation continues (**Figure 1**).

2. Urban river degradation

Full urban development then creates a disconnect between the benefits of urban infrastructure and the environmental costs needed to sustain it. A state of evolving equilibrium or permanent disequilibrium occurs generating pressing problems that become difficult to solve. These problems are becoming well studied and documented.

2.1 Erosion and sediment mobilization

Erosion is the fastest and most prominent visual effect of urbanization. Wolman [4] is credited with some of the earliest documentation of erosion and sedimentation directly attributed to the built environment. His observations documented export of sediment from forested watersheds (< 40 tons/year/Km^2) into agricultural (120-320 tons/year/Km^2) to urbanized/developed (>4,000 tons/year/Km^2). Further research into this problem yielded estimates from active construction as high as 50,000 tons/year/Km^2 [5]. The impact is diluted depending on the size of the watershed but nonetheless problematic. The final observation by Wolman [4] visually described his observations. Our urban river environment is a panorama of flood debris, sand, scoured bottoms and exposed sediment. Unfortunately, this still accurately describes the condition of many urban rivers today.

Further research and understanding of this paradigm were advanced by Chin [6]. She quantified that initial urban development mobilizes sediment on the magnitude of 2-10 times the natural rates. After development subsides, lower sediment yields predominate (< 30 tons/year/Km^2) but a new hydrology is established. A new and enlarged river channel 2 – 15 times the original size is needed to accommodate changing water volumes discharged directly to the stream. Even though the stream enlarged, erosion continued at a rate of about 0.3 meters per year [6]. Erosion will continue until the stream channel equilibrates to watershed disturbance or more likely never subsides due to a constant pattern of disequilibrium. Discolored urban rivers of brown or orange are typical during most rain events because of this problem.

Current analysis by Gregory [7] and Maklin and Lewin [8] suggest a more holistic paradigm for river change. Time has elapsed since early urbanization impacted river systems and researchers now incorporate time scales into thinking about human impact on these systems. Small scales and short time periods cannot encompass the entirety of impacts. River systems are variable and differential in response to perturbations. While some systems are resilient others are not. Some move into a sustainable pattern while other continue a disequilibrium continuously. All river systems are impacted and now researchers are working to quantify this impact.

Because precipitation is such a strong driver on these systems, as long as the natural pattern of infiltration has been disrupted we will continue to have problems. Some river systems are continually plagued by a dense blanket of eroded material while others suffer from highly erodible river banks. These problems began in the climatically benign twentieth century [8] but will need solved under changing climatic conditions and new precipitation patterns of the next century. This may prove more challenging or even catastrophic than anything we have previously faced. Rivers need to be understood from a global scale to within small reaches to encompass the entirety of change.

2.2 Changes in morphology and channelization

Beyond erosion, the continuing negative impact of a new hydrology on the physical (morphology) of the river environment is quite destructive. Excessive flow is disruptive and causes scouring of the stream bed, loss of habitat, streambank incision and isolation of the floodplain from the main channel [9]. High flow increases sheer stress on river bottom material scouring and pushing it downstream in what is called bedload. This scouring removes gravel and other material in the range of 2-64 mm [10] replacing it with bars and benches from material washed from erosional sediment above [6]. During dry periods, lack of infiltrating precipitation lowering the water table reduces lateral discharge. This creates periods of low flow degrading the aquatic environment further [11]. Rivers become 'flashy' suggesting periods of very high and very low flow rather than a consistent and stable hydrological regime.

Macroinvertebrate and fish habitat suffers. The good quality sediment and riffle habitat that the 2-64 mm material created is lost eliminating the essential refuges for aquatic life. Replacement sediment bars and benches subject to similar bed sheer stress [12] constantly change making this new environment unstable. Continual stream bank erosion adds more sediment to the stream channel until isolation from terrestrial interaction is completed [13]. Food webs become disconnected as we see less biodiversity and abundances of fish, aquatic life and the woody debris these problems generate [14].

Isolation from the floodplain is very problematic. Restriction of this hyporheic exchange limits nutrient reduction, temperature regulation and pollutant removal [15]. Rivers are part of a larger and interconnected system that cannot function well in isolation. Without the purifying mechanisms and ecological connectivity of the floodplain, the river is reduced to nothing more than a water conveyance.

2.3 Pollution problems compounded by sedimentation

Sediment mobilization from changing hydrology further complicates the pollution problems surrounding urban environments. Sediment is known to transport increased levels of many pollutants [16]. Influxes of plastics, debris and other solid waste flow freely into rivers because of direct stormwater and overland flow. Land in proximity to urban areas and watersheds with a high urban land use strongly correlate with increased concentrations of microplastics [17, 18]. Microplastics once mobilized into the river environment are concerning because of their ability to sorb/release persistent organic contaminants [19, 20]. These plastics also act to transport and provide surfaces for the growth of microbiological pathogens [21]. This complicates efforts to reduce the bacterial loading we find in these systems and heightens the concern for disease. Finally, the transport of these microplastics into the oceans from urban river drainage is very concerning and problematic [19, 22]. This problem must be controlled first in urban watersheds to provide any hope of reducing the impact in our oceans.

Pathogens (bacteria, protozoans, viruses) easily flow through the urban river environment entering from stormwater, wastewater and overflowing or leaks from sanitary sewer systems. Of these sources, stormwater generates the greatest impairment to urban rivers because of water volume [23] and concurrently is the greatest concern for disease outbreak. Using climate and epidemiological records, Rose et al. [24] found statistical evidence suggesting a correlation between storm events and disease outbreak in cities. Sediment loading of river beds along with organic material provides a good environment for bacteria such as *E. coli* to survive until the next storm event re-suspends them into river water. Pachepsky and Shelton [25] found the survivorship of *E. coli* in sediments was much greater than in overlying waters. Mallin et al. [26] attributed continual bacterial contamination from a sewage spill to release from underlying sediments well after levels depleted in overlying water. This creates the concern that urban rivers harbor extensive beds of bacteria, potential pathogens and other pollutants that will be resuspended continually as sediment and plastics move through these systems.

2.4 Flooding and impervious surfaces

Interwoven into all of these problems are the changes in flood periodicity and intensity. The urban drainage network influences the river flood regime from response time due to precipitation events through the ultimate magnitude of the flood [27]. The highly impervious urban watershed has a diminutive ability to minimize flooding generating ever increasing amounts of surface runoff [28].

Development acts to amplify the runoff response causing smaller and smaller precipitation events to generate larger and larger flood events. Rainfall intensity rather than duration then becomes the driving force behind urban flooding.

Researchers investigating this phenomenon began to characterize these patterns and search for solutions. Initial characterization suggested nonporous landscapes like parking lots and buildings behaved collectively as an impervious barrier to precipitation infiltration. Calculated as a percentage, increasing coverage corresponded directly with greater volume of stormwater discharge into a river without treatment. Researchers studied how these impervious surfaces operated then incorporated these ideas into a model of impervious cover [29]. The model suggests an increasing level of stream degradation corresponding to incremental thresholds of impervious surface. Increases up to 10% of impervious surface throughout a watershed cause the river to become sensitive to inputs. Between 10% and 25%, the river becomes impacted or impaired. Beyond 25% impervious cover, the river becomes non supporting of essential river functions.

Further research found that stormwater infrastructure is actually more predictive of stream degradation than percentages of impervious surfaces [30, 31]. Research suggests that the increased complexity of stormwater of pipes and drains, the greater the impact on receiving streams. Effective Impervious Surface (EIA) was developed as a better descriptor than Total Impervious Surface (TIA) when predicting river response [32]. EIA uses the connectivity of stormwater discharge directly into the river where TIA calculates only the total surface area. Schuster et al. [28] explains the problems associated with EIA. With just a 10% level of effective impervious surface (EIA), runoff production increases to the extent that 2-year intensity storm now yields the same amount of discharge to the river as a 10-year storm. This is profound because stormwater infrastructure has now fundamentally changed watershed function. Smith et al. [27] found that the five largest floods in past 74 years in Charlotte, North Carolina occurred after 1995 suggesting that the drainage density (EIA) created this response. Such conclusions are corroborated throughout the literature [33] generating concerns that urban watersheds fundamentally changed by EIA are ill equipped to protect streams and rivers from impending climate change.

2.5 Stormwater infrastructure

Most stormwater infrastructure was built around the central premise of peak attenuation. Development requires mitigation of excessive stormwater created by the impervious surface. Most often this is some form of detention pond or other structure to slow water flow into a receiving river. The theory behind these structures is to capture the newly created runoff from development, hold it in place and then later release it at rates no greater than the pre-development peak. Thus, the peak is shaved and flattened and theoretically mimics what was discharged before development. Ecologically, this theory is flawed because a new and different stormwater peak has been generated. Roesner et al. [34] reviewed why this is so damaging to the river environment. This practice exposes the stream to extended periods of flow rather than the previous slow infiltration and discharge vegetated watersheds provided. While the peak is shaved, a greater volume of surface runoff is created and receiving streams are not in equilibrium to receive it. Further, the one size fits all mentality of design ignores unique attributes of urban landscapes for expediency. Meeting only minimum regulatory requirements (usually no greater than a ten-year storm) has built a watershed landscape that is easily overwhelmed during high intensity precipitation. Intensity-duration-frequency (IDF) curves used to engineer stormwater infrastructure may have undersized the entirety of our urban landscape as climate change impacts future precipitation patterns [35].

Even more problematic, these structures condense what was once a diffuse overlay of precipitation throughout an area into a single point discharge. This has immense ramifications on the receiving stream channel. While these structures are capable of storage and attenuation of small storms, research suggests they are highly ineffective for larger storms [36]. Further, these structures are not protective of overall degradation and any protective capabilities reduce with age [37]. Stormwater infrastructure will be very problematic as we desire to improve watersheds in the future.

3. Monitoring water quality in urban Rivers

Good quantification of urban impacts on rivers is needed to determine levels of degradation and begin the rehabilitation process. Depending on instrumentation, budgets and personnel; many water quality parameters are available with various uses toward the prediction of water quality (**Table 1**). Decisions on what to measure rests behind the objectives of the study and translation into effective policy for improvement.

Decision making begins with methodology and site selection gaining good access to a river for study. Barbour et al. [38] outlines several approaches. A targeted study where concerns over a specific outfall or disturbance entering the river may be one approach. Here, samples are taken above, in and below the concern. Comparisons are made to determine the extent of impact. Another approach is collection of information to assess the overall condition of the river and watershed. Sites are selected throughout the river basin then compared to a reference condition or norms of water health. In severely degraded urban areas historical data may be necessary for comparisons.

Concerning seasonality, samples should be taken during each significant season then characterized. Based on findings, an index time period is created to meet objectives. This allows the investigator to collect data during that time period and interpret findings within the bounds of the study. This approach is good for an overall analysis of urban river quality. If specific outfalls or problem areas are the concern seasonality may not need analysis.

Methodologies are dependent upon the parameter used. Water samples are measured using field instrumentation such as submersible meters and laboratory analysis detailed in publications such as Standard Methods [39]. Insects are collected using various types of nets, preserved and later sorted then enumerated. A rapid bioassessment (RBA) technique developed by Barbour et al. [38] is possible or if a more detailed approach is needed the use of bottom samplers such as a Surber or Hess is warranted. Comparisons between RBA and Surber methods have been studied [40]. For fish collections, electrofishing is the preferred collection technique [41]. It describes the pulsing of DC current into water temporarily immobilizing fish allowing collection for analysis.

Chemical water quality parameters are used individually or compiled into a metric dependent upon the investigator and the study. More data strengthens conclusions, however any data collection derived through a targeted study is beneficial. In 1970, a group of 142 water quality scientists developed a Water Quality Index (WQI) [42]. Using 9 prominent parameters (dissolved oxygen, fecal coliform/*E. coli*, pH, biochemical oxygen demand (BOD) (5-day), temperature change (from 1 mile upstream), total phosphate, nitrate, turbidity and total solids) the index was created for use in defining water quality. Multiple variations of the WQI have evolved (reviewed by Bharti and Katyal [43]) and are effective within defined use. Further, Noori et al. [44] explored the substitution of alternative measures into the

Parameter	Description	Uses
Dissolved Oxygen	Concentration of oxygen dissolved in water.	Loss of oxygen suggests organic pollution and high BOD. Supersaturation suggests nutrient pollution and stimulation of plant growth.
Temperature	Heating and cooling of the water.	Urbanization creates heat islands warming river water.
pH	Changes in the hydrogen ion concentrations.	Rising pH suggests loss of organic material entering stream from deforestation. Runoff from impervious surfaces increases pH.
Conductivity	Measure of strength of electrical charge.	Direct correlation between increasing conductivity and pollutants. Sewage contains a very high conductivity.
Alkalinity	Buffering capacity.	Poorly buffered rivers have significant changes in pH.
ORP	Oxidations and Reduction Potential.	Healthy streams are well oxidized and pollution tends to drive ORP lower.
Turbidity	Discoloration of water.	Water increases in turbidity as it is polluted with sediment and other pollutants.
TS, TSS and TDS	Measure of solids in water.	Suspended solids are a form of pollution increasing as water degrades. Important to distinguish between Total Solids (TS), Total Suspended Solids (TSS) and Total Dissolved Solids (TDS) when making conclusions.
BOD	Biological Oxygen Demand.	Oxygen consumption by bacteria breaking down organic matter. Increases in BOD degrade water quality.
Fecal Coliforms/*E. coli*	Bacterial contamination and pathogen estimation.	Indicators of disease potential and increases suggest water of worsening quality. At certain levels of contamination water should be avoided.
Phosphorus	Concentration of limiting nutrient.	Believed to be a keystone pollutant due to very low concentrations in freshwater suggesting increases in this nutrient signal increases in many pollutants.
Nitrogen	Concentration of polluting nutrient.	Various species of nitrogen helpful in pinpointing types of pollution such as agricultural runoff or sewage contamination.
WQIs	Various indices predicting water quality using chemical parameters.	Compilation of many combinations of chemical measures to produce an overall measure of water quality. May be limited to short-term changes in river quality.

Parameter	Description	Uses
IBIs - Macros	Various indices predicting water quality using aquatic insect abundances and diversity.	Compilation of many combinations of macroinvertebrate assemblages (abundance, diversity and sensitivity) to produce an overall measure of water quality. Better measure as these insects integrate environmental change over longer time periods.
IBIs - Fish	Various indices predicting water quality using fish abundances and diversity.	Compilation of many combinations of fish assemblages (abundance, diversity and sensitivity) to produce an overall measure of water quality. Good measure as fish integrate environmental change but mobility and habitat destruction must be considered.
Remote Sensing	Estimates of water clarity, chlorophyll concentration, organic and mineral suspended material.	This technique uses the visible and near infrared light bands of the solar spectrum to predict water quality through correlations between the water column reflections and known concentrations of parameters measured. This technique is not adequate alone but very useful in coordination with other measured parameters.
Data Loggers	Dissolved Oxygen, Conductivity, Water Level and pH.	Data loggers allow deployment into water environments to capture continuous data for a particular parameter. These loggers are highly advantageous for long term data acquisition and areas hard to access. They need to be well moored to structures to avoid loss during extreme flooding events.
Water Quality Models	Most common include: AQUATOX, CE-QUAL-W2, Environmental Fluid Dynamics Code (EFDC), QUALs (QUAL2E, QUAL2E-UNCAS, QUAL2K, and QUAL2Kw), Soil and Water Assessment Tool (SWAT), Spatially Referenced Regression on Watershed Attributes (SPARROW), and Water Quality Analysis Simulation Program (WASP).	Models are used to assist in effective water quality management and to assist in decision making. Models are also useful in reducing the need for extensive sampling. It is important to note the region, precision and ecosystem specifics for each chosen model. Models are a good choice to assist with management decisions.

Table 1.
Water quality measures along with descriptions and uses.

original WQI. Changes in water quality classification were observed when other parameters are substituted so care must be exercised when calculating the WQI.

Overall, WQI is a good predictor of water quality and useful in river characterization. However, the ability to monitor all required parameters may be beyond the capabilities and means of a researcher. Kannel et al. [45] suggested a minimized WQI version using only 5 parameters (Temperature, pH, Dissolved Oxygen,

Conductivity and TSS) and an additional index using only dissolved oxygen as an effective alternative. While the minimized indices were not as predictive as the full WQI, they were good for periodic measures. This suggests that some level of river water quality can be monitored with minimal amounts of resources, time and effort.

Alternatively, the condition of the stream may be assessed by assembling collections of insects into an index. Many metrics have been identified and reviewed [46], so inclusion into your study depends on objectives. Most indices use abundance and sensitivity to pollution for water quality determinizations. When using insect indices, the collection method is very important. Available resources for collection, enumeration and taxonomic expertise all impact the metric used. These factors must be established initially to properly assign an index.

The Index of Biotic Integrity (IBI) categorizes the health of a stream using fish populations [47]. Modified versions of the IBI are used to reflect the condition of waters in various regions. Often, multiple metrics are used then summarized to develop an overall metric of water quality. Each metric is scored with a 5 if it reflects a system with very little human influence or a 1 if it departs significantly from a reference stream. A score of 3 is used to describe intermediate qualities. The index is effective yet the criteria for choosing different metrics and selection of reference conditions are issues that need addressed when using an IBI [48]. The index can be strengthened when used in conjunction with other water quality metrics.

Remote sensing, data loggers and modeling are additional methodologies used to quantify water quality. Each method contains limitations, advantages and variable cost so all factors must be weighed before incorporation into an assessment program. The advantages of remote sensing include collection capabilities over very broad areas and documenting comprehensive historical records of change [49]. Collected data can be used to prioritize where to concentrate localized sampling effort but any use must be complimented with ground level measures. It is important to incorporate this type of work with the other methodologies listed. Data loggers such as the HOBO [50] are useful for deployment into rivers to collect sets of continuous data. This is advantageous when monitoring oxygen or conductivity but is limited to these and a few other physical parameters. Water quality models have universal utility to aid decision making and management [51]. Model type, calibration and sensitivity must be carefully selected to ensure the best fit for the study.

4. Protections and management

The regulatory environment surrounding protections of stream ecosystems is vast. All are designed to understand the pollutant loading to the streams, create a plan for minimization and then implement these technologies throughout the built environment. Each addresses different parts of the problem such as point and non-point sources of water pollution.

4.1 Point pollution control

Point source water pollution control occurs throughout the world. In Europe, a directive designed for member states to collect and treat urban wastewater went into effect in 1991 [52]. Freedom was given each state to meet requirements based on reduction goals and receiving water classifications. In the USA, the Environmental Protection Agency (EPA) established the National Pollutant Discharge Elimination Permit (NPDES) program as part of the 1971 Clean Water Act to control any facility discharging waste directly into a stream [53]. In other parts of the world, point source discharge is varied. In China, worsening water quality throughout the major water basins has generated national pollution discharge standards [54]. These standards are technology driven and challenges remain with implementation at the local level. Other parts of the world may be considered even less centrally regulated. While Latin America is known for a system of concentrated central government, water protection law is often spread throughout multiple disparate agencies and ununiformly enforced. Shahady and Boniface [55] reviewed water law in Costa Rica and found five distinct agencies throughout health, sanitation, irrigation, regulation and local water boards with shared responsibility in water protection. This system is very effective in some areas while lacking in others.

This permitting system has shown success toward improving water quality. In the USA, direct reductions in pollutants from wastewater have shown water quality improvements even in large systems such as the Chesapeake Bay [56]. In Europe, clear progress has been made in reducing emissions into urban rivers and lakes; this has been done through connections to sewers, the introduction of wastewater treatment and the upgrading of earlier treatment plants [57]. Elsewhere challenges remain. In much of the developing world urbanization is proceeding at a much faster rate than treatment infrastructure can accommodate [58]. There is room for considerable water quality improvement through point source control in many areas of the world [59].

4.2 Non-point source control

In the USA, the Total Maximum Daily Load (TMDL) program is the current Stormwater Management (SWM) methodology used by the EPA to protect receiving water systems from stormwater pollution. A TMDL calculates the maximum amount of a pollutant that a water body can receive and still meet water quality standards. Once that calculation is made, pollutant allocations (point and non-point) are partitioned throughout all sectors allowing only enough pollutants into the river to maintain identified use. A similar program exists in Europe. River Basin Management Plans (RBMPs) are required documenting impairment from diffuse (non-point) sources and plans for improvement. The idea of pollutant loading identification and then curtailing influx to the river is the consensus for controlling this type of pollution. In fact, some level of SWM is considered an important environmental issue in countries worldwide regardless of their level of development [60].

Successful implementation of SWM requires installation of some type of green infrastructure, changes in land management or permit modifications. Other tools available include mitigation banking [61] and nutrient trading [62]. These tools work in concert with other options to create an economic blueprint to fund the needed reductions to meet SWM requirements.

4.3 BMPs and other protections

For non-point source control, the construction of some type of infrastructure is required due to the diffuse nature of the pollutant. This field has taken on a wide variety of terms [63]. Low Impact Development (LID) is construction that attempts to mimic the natural hydrology and so encompasses any installation that reduces stormwater impact. Water Sensitive Urban Design (WSUD) is a similar approach and includes any effort to minimize the hydrological impacts of urban development on the surrounding environment. Integrated urban water management (IUWM) is a somewhat broader concept combining the management of water supply, groundwater, wastewater and stormwater. Sustainable Urban

Drainage Systems (SUDS) consist of a range of technologies and techniques used to drain stormwater/surface water in a manner that is (arguably) more sustainable than conventional solutions. Best Management Practices (BMPs) are those practices that possess both non-structural (street sweeping) and structural (retention pond) attributes that minimize impact of stormwater. Stormwater Control Measures (SCMs) are identical but this term is used to eliminate the idea of best because alternative practices (not the best ones) can be used. Alternative techniques (ATs) or compensatory techniques (CTs) describe structures used to reduce runoff volume, peak flows and flooding. Some of these techniques can be considered for the protection of the quality of receiving environments. Source Control (SC) is used for on-site stormwater systems. Green Infrastructure (GI) is the part of urban planning that utilizes green space hubs and corridors highlighting their potential ecosystem services.

All technologies are designed to provide benefits to the urban environment and to mitigate the harmful effects of stormwater. The effectiveness of these technologies is dependent on soil characteristics, proper design and installation. Many may be limited by the volume of water flowing into them as retention time is a critical component for effective treatment. Soil is another critical component. A soil with a slow percolation rate cannot handle the volume of water that a good percolating soil is able to infiltrate. The percolation of soil must also be considered in flooding risk. As soil moisture exceeds 45%, pervious areas generate runoff contributing directly to stream discharge. Thus, green infrastructure is now understood to underperform in large and high intensity storm events resulting in flooding [64].

Retention or detention basins and small ponds are the most common technologies installed due to cost and ease of construction. These basins meet multiple design criteria from peak shaving that is required for any land disturbing activity to infiltration by pooling water allowing it to infiltrate. Retention basis (ponds) also provide habitat and recreational opportunities making them appealing in communities. These ponds (<0.01 Km^2) may be responsible for 34% of the nitrogen, 69% of the phosphorus and 12% of the sediment masses retained collectively by all aquatic components in the watershed [65]. More study is needed to quantify the collective impact the network of these small ponds has on urban watersheds.

We now have an ability to create a water sensitive city [66]. This includes the ubiquitous use of plants in any design to create the potential for removal nutrients and other pollutants. Green roofs installed on rooftops now intercept precipitation reliving some of the burden on the stormwater infrastructure and the need for peak attenuation. Building blue roofs to create water storage keeps water out of the stormwater cycle easing the burden on stormwater infrastructure. These alternatives provide a visual improvement over a conventional roof top and a return on investment by lowering environmental costs. They also help to regulate building temperatures improving energy efficiencies. Green design forces water discharge from roads into some form of bioretention further improving water quality [67, 68]. With these designs, stormwater is still flowing but at a reduced rate and with better quality.

Wetlands with or without larger impoundments add water quality treatment, support a diversity of wildlife and provide recreational opportunities within urban environments. The key component of these designs is retention of water in the system. Entering water is spread out evenly and soils constantly inundated to produce the reduced water conditions needed for a wetland. Wetlands can improve stormwater considerably by absorbing flow and mitigating concentrations of nutrients and bacteria [69]. Recreational boardwalks and interpretation signage are additional features these environments provide.

4.4 Stream restoration

In many instances, pollutant source reductions and built SWM infrastructure may not be enough to recapture watershed integrity because of stream channel alteration. In these instances, restoring stream channels is now a preferred goal. In these designs, hydrology, sediment transport and watershed processes that have fundamentally changed are incorporated into a new channel design [70]. In these designs, streambank stabilization, restoration of the stream channel and reestablishment of riparian vegetation are the areas of focus. To stabilize banks, jetties, tree revetments, rock vanes, rock toes, retaining walls and gravel banks are used [71]. While retaining walls and gravel banks provide good stability they are expensive. Jetties may not provide the same level of protection but are the most cost effective measure and provide a more natural look to the restoration. Tradeoffs exist between appearance, effective stabilization and costs.

To rehabilitate the stream channel, regional curve dimensions, planform pattern, and grade control structures are developed [72]. Grade control structures hold the vertical elevation of the stream constant preventing vertical downcutting. Planform pattern creates an alignment of the stream channel to resemble meanders typical for the regional landscape. Proper channel width and depth are created from a "best-fit" consideration using the bankfull channel dimensions of similar reference stream from within the drainage area.

Unfortunately, there is considerable scientific evidence that instream restoration in USA and Europe has shown very limited success [73–77]. Storm events that exceed bankfull have fundamentally changed restoration design reducing expected performance [78]. Restoration of features (stormwater ponds, riparian vegetation) outside the stream channel have shown some improvement but well under expectations based on investment [79]. Prevention remains the single most effective

Figure 2.
Channelized urban river in need of stream restoration.

restoration technique. Lack of effective measure for restoration may be sensitive to time scales as full recovery may take over 15 years to be realized [80].

The how and where to restore is difficult to decide. Urban land is expensive and the best areas may be privately owned and unwelcoming. Projects are driven by available land to municipalities and public sentiments rather than effectiveness in bringing about the best project restoration outcomes [81]. So because of the expense and limited measured effectiveness restoration projects are currently under debate. It appears that good channel restoration projects are best when integrated into multi management efforts including protection of existing good quality stream sections, reducing stormwater flow, controlling sewage overflows upgrading sewage treatment facilities (**Figure 2**) [82].

4.5 Sediment management

Managing sediment in urban stream channels may be paramount for improving urban river systems. Sediment moves through these systems in what can be characterized as the "urban sediment cascade" [83]. In this cascade, sediment is generated from two primary sources; first roads and impervious surfaces and then the bed of the aquatic system. As the sediment flows through this urban cascade it mixes with multiple contaminants such as metals [84] and microplastics [85]. This sediment flows from parts of the urban landscape such as street surfaces, pot holes, storm sewers, ditches and docks eventually entering rivers and lakes. This phenomenon makes every storm event a polluting event with the concentration of contaminates dependent upon the storm intensity.

Once in the stream bed this sediment may accumulate for extended periods of time. Evidence suggests movement of sediment fractions larger than the median size of the bed surface material is rare and occurs only at relatively high flows [86]. Such flows may occur once every few years and the movement might not last more than a few hours. This further suggests years of accumulation of contaminated sediment may be severe and pose a possible health risk [87]. Removal of dams as part of an overall restoration strategy to improve fish passage and sediment flow downstream may in actuality be counterproductive toward restoration goals as beneficial evidence of this practice is highly experimental [88, 89]. Prevention or clean up through practices such as street sweeping may be a better management strategy [90].

4.6 Citizen science and education

The importance of citizen science protecting water quality [91] is becoming more widespread and may be integral to restoring these systems. It bridges the gap between regulators and the public energizing citizens living in urban areas impacted by poor river quality. Citizens can be trained to complete essential tasks, are affordable and can generate good data when verified [92]. Programs such as the The Izaak Walton League's Save Our Streams [93] builds an army of volunteers through training and information to monitor our waterways. This program serves as an intermediary successfully uncovering problems and urging local leaders to take action. These types of programs can even transcend data collection and scientific analysis moving those involved toward a greater sense of place in the watershed [94]. This can further the idea of watershed protection and lead to real policy change.

Education is the other effort underway to secure river protection. Theoretically, an educated public will hold regulators and developers responsible for their actions. Programs such as Global Rivers Environmental Education Network (GREEN) educates global citizens about water quality problems [95]. Education based curriculum teaching stormwater principles in schools educates future generations about the problems rivers are facing [96]. Such programs are productive. But education is complex. People formulate perceptions about the environment using various levels of experience, normative influences, temporal discrepancies and attitude-behavior measures [97]. Why should people care or why be involved? Answers to these questions are extremely diverse, complex, and poorly understood. The furtherance of educational understanding can only enhance opportunities for improvement.

5. Conclusions

There is general agreement that urban water resources need to be managed better. We all depend on clean water for survival and the make-up of our communities reflects the quality of water resources flowing through them. The need for improvement is great and good policy needs both the natural and social sciences to generate good governance of our water resources [98]. But urban water management seems stuck in a state of maladaptation essentially locked into a societal need for large infrastructure and the never ending need to maintain it [99]. Change only occurs in response to some catastrophic or shock event (such as a flood) that in no way bears the needed one-on-one logical change to what triggered the shock event in the first place. This leads to changes in urban infrastructure that may bear no resemblance to real societal needs or watershed management.

What is ultimately needed is water policy adapted to societal needs instead of knee jerk responses to a crisis. Resources spent on stream restoration and flood mitigation upstream will be well received when urban residents understand the benefits downstream [100]. Local efforts to pick up trash and minimize plastics may have far reaching impact on communities when understood in the context of the world's oceans [101]. Or the necessity to rebuild current infrastructure to deal with ongoing climate driven precipitation change [102]. Good monitoring can expose the need and document the improvement. Urban rivers can be rehabilitated given a dedicated citizenry aided by governmental and scientific support. Our future depends on it.

Acknowledgements

The author would like to acknowledge the University of Lynchburg for providing the time and resources necessary to complete this project. Countless students have assisted me in my work to quantify the impact of urbanization on river environments and while they cannot be listed by name here I am appreciative. Other projects including work with Leesville Lake Association and the Sustainability Center in Costa Rica further inspired this work. My family is always at the forefront of support and no project can be completed without them. No external funding was provided for the completion of this project.

Author details

Thomas D. Shahady
University of Lynchburg, Lynchburg, Virginia, USA

*Address all correspondence to: shahady@lynchburg.edu

References

[1] Leopold L, Huppman R, Miller A. Geomorphic effects of urbanization in forty-one years of observation. Proceedings of the American Philosophical Society. 2005; 149: 349-371.

[2] Walsh C, Roy A, Feminella J, Cottingham P, Groffman P, Morgan II R. The urban stream syndrome: Current knowledge and the search for a cure. Journal of the North American Benthological Society. 2005; 24: 706-723.

[3] Meyer J, Paul M, Taulbee W. Stream ecosystem function in urbanizing landscapes. Journal of the North American Benthological Society. 2005; 24: 602-612.

[4] Wolman M. A cycle of sedimentation and erosion in urban river channels. Geografiska Annaler. 1967; 49: 385-395.

[5] Piest, R, Miller C. Sediment sources and sediment yields. In: Vanoni V, Editors. Sedimentation Engineering. American Society of Civil Engineers; 1975. p. 437-493.

[6] Chin A. Urban transformation of river landscapes in a global context. Geomorphology. 2006; 79: 460-487.

[7] Gregory G. Human influence on the morphological adjustment of river channels: The evolution of pertinent concepts in river science. River Res Applic. 2019; 8: 1097-1106.

[8] Maklin M, Lewin J. River stresses in anthropogenic times: Large-scale global patterns and extended environmental timelines. Progress in Physical Geography. 2019; 43: 3-23.

[9] Anim, D, Fletcher T, Vietz G, Pasternack G, Burns M. Effect of urbanization on stream hydraulics. River Res Applic. 2018; 34: 661-674.

[10] Pizzuto J, Hession W, McBride M. Comparing gravel-bed rivers in paired urban and rural catchments of southeastern Pennsylvania. Geology. 2000; 28: 79-82.

[11] Hardison E, O'Driscoll M, DeLoatch J, Howard R, Brinson M. Urban land use, channel incision, and water table decline along coastal plain streams, North Carolina. J. Am. Water Resour. Assoc. 2009; 45: 1032-1046.

[12] Morley S, Karr J. Assessing and restoring the health of urban streams in the Puget Sound basin. Conserv. Biol. 2002; 16: 1498-1509.

[13] Booth D, Karr J, Shauman S, Konrad C, Morley S, Larson M, Burges S. Reviving urban streams: land use, hydrology, biology, and human behavior. J. American Water Works Assn. 2004; 40: 351-364.

[14] Elosegi A, Díez J, Mutz M. Effects of hydromorphological integrity on biodiversity and functioning of river ecosystems. Hydrobiologia. 2010; 657: 199-215.

[15] Driscoll M, Clinton S, Jefferson A, Manda A, McMillan S. Urbanization effects on watershed hydrology and in-stream processes in the southern United States. J. Austr. Water Wastewater Assoc. 2010; 2: 605-648.

[16] Mallin M., Johnson V, Ensign S. Comparative impacts of stormwater runoff on water quality of an urban, a suburban, and a rural stream. Environ Monit Assess. 2009;159:475-491.

[17] Baldwin A, Corsi S, Mason S. Plastic debris in 29 Great Lakes tributaries: relations to watershed attributes and hydrology. Environ. Sci. Technol. 2016;50:10377-10385.

[18] Mani T, Hauk A, Walter U, Burkhardt-Holm P. Microplastics profile

along the Rhine River. Sci. Rep. 2015;5:17988.

[19] Andrady A. The plastic in microplastics: a review. Mar. Pollut. Bull. 2015;119:12-22.

[20] Wang F, Wong C, Chen D, Lu X, Wang F, and Zeng E. Interaction of toxic chemicals with microplastics: a critical review. Water Res. 2018;139:208-219.

[21] McCormick A, Hoellein T, London M, Hittie J, Scott J, Kelly J. Microplastic in surface waters of urban rivers: concentration, sources, and associated bacterial assemblages. Ecosphere. 2016;7:1-22.

[22] Wijnen J, Ragas M, Kroeze C. Modelling global river export of microplastics to the marine environment: Sources and future trends. Science of the Total Environment. 2019;673:392-401.

[23] Arnone R, Walling J. Waterborne pathogens in urban watersheds. J. Water Health. 2007;5:149–162.

[24] Rose J, Daeschner S, Easterling D, Curriero F, Lele S, Patz J. Climate and waterborne disease outbreaks. J. Am. Wat. W. Assoc. 2000; 92:77-87.

[25] Pachepsky Y, Shelton D. Escherichia coli and fecal coliforms in freshwater and estuarine sediments. Crit Rev Environ Sci Technol. 2011;41:1067-1110.

[26] Mallin, M, Cahoon L, Toothman B, ParsonsD , McIver M, Ortwine M, Harrington R. Impacts of a raw sewage spill on water and sediment quality in an urbanized estuary. Marine Pollution Bulletin. 2007;54:81-88.

[27] Smith J, Baeck A, Morrison M, Sturdevant-Rees J, Turner-Gillespie P, Bates P. The regional hydrology of extreme floods in an urbanizing drainage basin. Journal of Hydrometeorology. 2002;3:267-282.

[28] Schuster W, Bonta J, Thurston H, Warnemuende E, Smith D. Impacts of impervious surface on watershed hydrology: A review. Urban Water Journal. 2005; 2:263-275.

[29] Schueler T, Fraley-McNeal L, Cappiella K. Is impervious cover still important? Review of recent research. Journal of Hydrologic Engineering. 2009;14:309-315.

[30] Walsh C, Fletcher T, Burns M. Urban stormwater run-off: a new class of environmental flow problem. PLOS ONE 2012;7:e45814.

[31] Ress L, Hung C, James L. Impacts of urban drainage systems on stormwater hydrology: Rocky Branch Watershed, Columbia, South Carolina. J. Flood Risk Management. 2020;13:e12643.

[32] Miller J, Kim H, Kjeldsen T, Packman J, Grebby S, Dearden R. Assessing the impact of urbanization on storm runoff in a peri-urban catchment using historical change in impervious cover. Journal of Hydrology. 2014;515:59-70.

[33] Sohn W, Kim J, Li M, Brown R, Jaber F. How does increasing impervious surfaces affect urban flooding in response to climate variability? Ecological Indicators. 2020;118: 106774.

[34] Roesner L, Bledsoe B, Brashear R. Are best practices management criteria really environmentally friendly? ASCE Journal of Water Resources Planning and Management. 2001;27:150-154.

[35] Kharin V, Zwiers F, Zhang X, Hegerl G. Changes in temperature and precipitation extremes in the IPCC ensemble of global coupled model simulations. Journal of Climate. 2007;20:1419-1444.

[36] Fry T , Maxwell R. Evaluation of distributed BMPs in an urban watershed—High resolution modeling

for stormwater management. Hydrological Processes. 2017;31:2700-2712.

[37] Moore T, Rodak C, Ahmed F, Vogel J. Urban stormwater characterization, control and treatment. Water Environment Research. 2017; 90:1821-1871.

[38] Barbour M, Gerritsen J, Snyder B, Stribling J. Rapid bioassessment for use in streams and wadeable rivers: periphyton, benthic macroinvertebrates and fish. 2ND ed. U.S. Environmental Protection Agency; Office of Water. 1999; EPA 841-B-99-002.

[39] Rice E, Baird R, Eaton A. Standard methods for the examination of water and wastewater. American Water Works Association/American Public Works Association/Water Environment Federation; 2017.

[40] Gillies C, Hose G, Turak E. What do qualitative rapid assessment collections of macroinvertebrates represent? A comparison with extensive quantitative sampling. Environ. Monit. Assess. 2009;149:99-112.

[41] Bohlin T, Hamrin S, Heggberget T, Rasmussen G, Saltveit S. Electrofishing - Theory and practice with special emphasis on salmonids. Hydrobiologia. 1989;173: 9-43.

[42] Brown R, McClelland N, Deininger R, Tozer R. A water quality index: Do we dare? Water and Sewage Works. 1970;117:339-343.

[43] BhartI N, Katyal D. Water quality indices used for surface water vulnerability assessment. Int. J. Environ. Sci. 2001;2:154e173.

[44] Noori R, Berndtsson R, Hosseinzadeh M, Adamowski J, Abyaneh M. A critical review on the application of the national sanitation foundation water quality index. Environ. Pollut. 2019;244:575-587.

[45] Kannel P, Lee S, Lee Y, Kanel S, Khan S. Application of Water Quality Indices and Dissolved Oxygen as Indicators for River Water Classification and Urban Impact Assessment. Environ Monit. Assess. 2007;132:93-110.

[46] Herman M, Nejadhashemi A. A review of macroinvertebrate- and fish-based stream health indices. Ecohydrology & Hydrobiology. 2015;15:53-67.

[47] Karr J. Assessment of biotic integrity using fish communities. Fisheries. 1981;6:21-27.

[48] Ruaro R. A scientometric assessment of 30 years of the index of biotic integrity in aquatic ecosystems: applications and main flaws. Ecol. Indic. 2013;29:105-110.

[49] Gholizadeh M, Melesse A, Reddi L. A comprehensive review on water quality parameters estimation using remote sensing techniques. Sensors. 2016;16:1298.

[50] Schmidt W, Raymond D, Parish D, Ashton I, Miller P, Campos C, Shutler J. Design and operation of a low-cost and compact autonomous buoy system for use in coastal aquaculture and water quality monitoring. Aquacultural Engineering. 2018;80:28-36.

[51] Costa B, da Silva C, Marques L, Almeida A, Leite I, de Almeida I. Applicability of Water Quality Models Around the World-a Review. Environmental science and pollution research international. 2019; 26:36141-36162.

[52] Directive EU. Council Directive of 21. May 1991 concerning urban waste water treatment (91/271/EEC). J. Eur. Commun. 1991;34:40.

[53] USEPA (United States Environmental Protection Agency). National Pollutant Discharge

Elimination System [Internet]. 2020. Available from https://www.epa.gov/npdes [Accessed: 2021-03-02].

[54] Li W, Sheng G, Zeng R, Liu X, Yu H. China's wastewater discharge standards in urbanization Evolution, challenges and implications. Environ Sci Pollut Res. 2012;19:1422-1431.

[55] Shahady T, Boniface H. Water quality management thorough community engagement in Costa Rica. Journal of Environmental Science and Studies. 2018;8:488-502.

[56] Lyerly C, Hernández Cordero A, Foreman K, Phillips S, Dennison W. (eds.). Lessons from Chesapeake Bay Restoration Efforts: Understanding the role of nutrient reduction activities in improving water quality. 2013.

[57] European Environment Agency. Rivers and Lakes in European Cities: Past and future challenges. 2016; EEA Report No 26/2016.

[58] Xu Z, Xu J, Yin H, Jin W, Li H, He Z. Urban river pollution control in developing countries. Nature Sustainability. 2019;2:158-160.

[59] Bernhardt E, Band L, Walsh C, Berke P. Understanding, managing, and minimizing urban impacts on surface water nitrogen loading. Annals of the New York Academy of Sciences. 2008;1134:61-96.

[60] Marsalek J, Chocat B. International report: Stormwater Management. Water Science and Technology. 2002;46:1-17.

[61] Lave R. Stream mitigation banking. Water. 2018;5:e1279.

[62] Hennessy R. Nutrient trading to improve water quality. Pollution Engineering. 2001;33:10-14.

[63] Fletcher T, Shuster W, Hunt W, Ashley R, Butler D, Arthur S, Trowsdale S, Barraud S, Semadeni-Davies A, Bertrand-Krajewski J, Mikkelsen P, Rivard G, Uhl M, Dagenais M, Viklander M. SUDS, LID, BMPs, WSUD and more – The evolution and application of terminology surrounding urban drainage. Urban Water Journal. 2015;12:525-542.

[64] Sohn W, Kim J, Li M, Brown R. The influence of climate on the effectiveness of low impact development: A systematic review. J. Environ. Manage. 2019;236:365-379.

[65] Schmadel N, Harvey J, Schwarz G, Alexander R, Gomez-Velez J, Scott D, Ator S. Small ponds in headwater catchments are a dominant influence on regional nutrient and sediment budgets. Geophysical Research Letters. 2019;46:9669-9677.

[66] Grant G. The Water Sensitive City . Chichester: Wiley; 2016.

[67] Li J, Deng C, Li Y, Song J. Comprehensive benefit evaluation system for low-impact development of urban stormwater management measures. Water Resour. Manag. 2017;31:4745-4758.

[68] Petit-Boix A, Sevigné-Itoiz E, Rojas Gutierrez L, Barbassa A, Josa A, Rieradevall J, Gabarrell X. Floods and consequential life cycle assessment: integrating flood damage into the environmental assessment of stormwater Best Management Practices. J. Clean. Prod. 2017;162:601-608.

[69] Mallin M, McAuliffe J, McIver M, Mayes D, Hanson M. High Pollutant Removal Efficacy of a Large Constructed Wetland Leads to Receiving Stream Improvements. Journal of environmental quality. 2012; 41:2046-2055.

[70] Niezgoda S, Johnson P. Improving the urban stream restoration effort:

identifying critical form and processes relationships. Environmental Management. 2005;35:579-592.

[71] Naisargi D, Mittelstet A. Quantifying Effectiveness of Streambank Stabilization Practices on Cedar River, Nebraska. Water. 2017;9:930.

[72] Hawley R. Making Stream Restoration More Sustainable: A Geomorphically, Ecologically, and Socioeconomically Principled Approach to Bridge the Practice with the Science. Bioscience. 2018;68:517-528.

[73] Miller S, Budy W, Schmidt J. Quantifying macroinvertebrate responses to in-stream habitat restoration: applications of meta-analyses to river restoration. Restoration Ecology. 2010;18:8-19.

[74] Palmer M, Menninger H, Bernhardt E. River restoration, habitat heterogeneity and biodiversity: a failure of theory or practice? Freshwater Biology. 2010;55:205-222.

[75] Louhi P, Mykrä H, Paavola R, Huusko A, Vehanen T, Mäki-Petäys A, Muotk T. Twenty years of stream restoration in Finland: little response by benthic macroinvertebrate communities. Ecological Applications. 2011;21:1950-1961.

[76] Sundermann A, Stoll S, Haase P. River restoration success depends on the species pool of the immediate surroundings. Ecological Applications. 2011;21:1962-1971.

[77] Ernst A, Warren D, Baldigo B. Natural-channel-design restorations that changed geomorphology have little effect on macroinvertebrate communities in headwater streams. Restoration Ecology. 2012;20:532-540.

[78] Smith S, Prestegaard K. Hydraulic performance of a morphology-based stream channel design. Water Resources Research. 2005;41:W11413.

[79] Smucker N, Detenbeck N. Meta-Analysis of Lost Ecosystem Attributes in Urban Streams and the Effectiveness of Out-of-Channel Management Practices. Restoration Ecology. 2014;22:741-748.

[80] Dyste J, Valett H. Assessing stream channel restoration: the phased recovery framework. Restoration Ecology. 2019;27:850-861.

[81] Bernhardt E, Sudduth E, Palmer M, Allan J, Meyer J, Alexander G, Follastad-Shah J, Hassett B, Jenkinson R, Lave R, Rumps J, Pagano L. Restoring rivers one reach at a time: Results from a survey of U.S. river restoration practitioners. Restoration Ecology. 2007;15:482-493.

[82] Bernhardt E, Palmer M. Restoring streams in an urbanizing world. Freshwater Biology. 2007;52:738-751.

[83] Taylor K, Owens P. Sediments in urban river basins: a review of sediment contaminant dynamics in an environmental system conditioned by human activities. J. Soils Sediments. 2009;9:281-303.

[84] Aung P, Mao Y, Hu T, Qi S, Tian Q, Chen Z, Xing X. Metal concentrations and pollution assessment in bottom sediments from Inle Lake, Myanmar. J Geochem Explor. 2019;207:106357.

[85] Watkins L, McGrattan S, Sullivan P, Walter M. The effect of dams on river transport of microplastic pollution. Science of the Total Environment. 2019;664:834-840.

[86] Church M, Hassan M. Mobility of bed material in Harris Creek. Water Resources Research. 2002;38:1237.

[87] Heise S, and Förstner U. Risks from historical contaminated sediments in

the Rhine basin. Water Air Soil Pollut Focus. 2006;6:261-272.

[88] Wohl E, Lane S, Wilcox A. The science and practice of river restoration. Water Resour. Res. 2015;51:5974-5997.

[89] Shahady T, Cleary W. The Importance of a Low-Head Dam Mitigating Water Quality Degradation in an Urbanizing Watershed. Journal of Environmental Management.

[90] Irvine K, Perrelli M, Ngoen-klan R, Droppo I. Metal levels in street sediment from an industrial city: spatial trends, chemical fractionation, and management implications. J. Soils Sediments. 2009;9:328-341.

[91] Capdevila A, Kokimova A, Sinha Ray S, Avellan T, Kim J, Kirschke S. Success factors for citizen science projects in water quality monitoring. Science of the Total Environment. 2020;728:137843.

[92] Cohn J. Citizen Science: Can Volunteers Do Real Research? BioScience. 2008;58:192-197.

[93] Middleton J. The Stream Doctor Project: Community-Driven Stream Restoration. BioScience. 2001; 51.4: 293-296.

[94] Church S, Payne L, Peel S, Prokopy L. Beyond water data: benefits to volunteers and to local water from a citizen science program. Journal of Environmental Planning and Management. 2019;62:306-326.

[95] Stapp, W. Watershed Education for Sustainable Development. Journal of Science Education and Technology. 2000;9:183-197.

[96] Sakowich C, Hull M, Malafronte S, Nassar V. Stormwater Awareness Through Youth Watershed Education [Internet]. 2018. Available from: https://digitalcommons.wpi.edu/iqp-all/5207 [Accessed 2021/02/02].

[97] Kollmuss A, Agyeman J. Mind the gap: Why do people act environmentally and what are the barriers to pro-environmental behavior? Environmental Education Research. 2002;8:239-260.

[98] Olsson L, Head B. Urban water governance in times of multiple stressors: an editorial. Ecology and Society. 2015;20:27.

[99] Huitema D, and Meijerink S. Realizing water transitions: the role of policy entrepreneurs in water policy change. *Ecology and Society. 2010;*15: 26.

[100] Hong C, Chang H. Residents' perception of flood risk and urban stream restoration using multi-criteria decision analysis. River Res. Applic. 2020;36:2078-2088.

[101] Amaral-Zettler L, Zettler E, Slikas B, Boyd G, Melvin D, Morrall C, Proskurowski G, Mincer T. The biogeography of the Plastisphere: implications for policy. Front. Ecol. Environ. 2015;13:541-546.

[102] Wijesiri B, Liu A. Goonetilleke A. Impact of global warming on urban stormwater quality: From the perspective of an alternative water resource. Journal of Cleaner Production. 2020;262:121330.

Chapter 6

Effects of Illegal Artisanal Gold Mining Operations on Groundwater Quality in Ghana: The Case of Ahafo-Ano South District

Maxwell Anim-Gyampo, Belinda Seyram Berdie, Enoch Bessah and Sampson Kwaku Agodzo

Abstract

The general properties and overall chemical quality for potability of groundwater in Ahafo-Ano South District of the Ashanti Region of Ghana have been evaluated. With respect to pH, about 92% of groundwaters were potable while 8% were acidic and not potable. Approximately 4%, 32%, 56% and 8% of sampled groundwater were soft, moderately soft, hard and very hard respectively. The overall chemical quality analysis of groundwaters showed that 20%, 28%, 40%, 4% and 8% had excellent, good, poor, very poor and unsuitable drinking water qualities respectively. Approximately 12%, 40% and 84% of As, Ni and Pb exceeded their respective WHO limits while 32% of Cd and Fe exceeded their respective limits for potable water. These higher concentrations of heavy metals were observed to have occurred in communities with intensive illegal gold mining operations. Inhabitants in these areas could potentially be more predisposed to potential health hazards including cancer, nervous system damage, low IQ in children, reduced growth of foetus and premature birth in pregnant women, and kidney damage. It is expected that illegal artisanal gold mining activities will be banned while policies aimed at providing alternative livelihood be instituted to minimize any potential health hazards on humans in the District.

Keywords: Illegal artisanal gold mining, groundwater, quality, Ahafo-Ano District, Ghana

1. Introduction

Water from surficial and underground sources is critical for the sustenance of all life forms on earth. Surface water is arguably the most easily available and well-harnessed drinking water source since creation. In tropical rainforest regions, surface water resources are the utilized and source of water supply to meet the demands of domestic, industrial and agricultural purposes. However, certain human activities and natural processes may alter the surface water quality, thereby limiting its ease and scope of reliability and usage. Such natural processes may include earthquakes, volcanic eruptions, landslides and hurricanes, and when such events are occurring at local and regional scales, they may alter the quality of surface water sources. Anthropogenically, the progressive industrialization and application of scientific methods in improving various livelihood demands by humans may result in the generation of varied amounts of wastes of varying toxicities, which when not properly handled may serve as potential sources of pollution to the environment including surface and groundwater resources.

One key activity of man that contributes to the pollution of drinking water sources is illegal artisanal gold mining activities. Globally, approximately 100 million people are considered to be directly or indirectly involved in such operations for their livelihood [1]. In Ghana, however, artisanal gold mining is estimated to have begun several decades ago with approximately one million people directly or indirectly involved [2]. The areas of operations are generally, rural settings with few peri-urban centers and a much fewer large urban centers that depend on both surface and groundwater resources for domestic, agricultural and industrial purposes. The rapid growth in urban and peri-urban population, coupled with the ever-increasing unemployment have resulted in the greater proportion of local and migrant youth (male and female) in Ghana, Burkina Faso, Mali Niger as well as Chinese nationals, getting in involved in illegal artisanal gold mining.

Despite the positivity of artisanal gold mining to the socio-economic development of societies, especially in most developing countries like Ghana [3], illegal artisanal gold mining activities utilizes crude and rudimentary techniques and environmentally unfriendly extractive processes. Examples may include but are not limited to the extraction of shallow alluvial gold deposits; diversion of streams and river courses to expose otherwise inundated beds for excavation; washing of alluvial ores at close proximities and many cases, inside river and stream channels. These may contribute to land degradation and water quality deterioration among others [4]. Furthermore, the utilization of mercury and cyanide to leach and/or amalgamate the gold during extraction, have been found to contribute to the release of harmful metals to the environment [5, 6]. These metals when found above permissible limits in drinking water and/or the food chain could be detrimental to human health and sustenance of the ecosystem [7].

According to [8], there is a great national concern with regards to the increasing activities of environmentally unfriendly mining activities in Ghana. The Ahafo-Ano South District in the Ashanti Region of Ghana is a predominantly rural area with few peri-urban and urban dwellings, where illegal artisanal gold mining activities occur. This has led to an influx of various people majority of which are youth from different parts of Ghana. Groundwater per the national water policy of Ghana [9] is the major source of potable water in rural Ghana. The observed rapid influx of people into the area to partake in the illegal artisanal gold mining requires that large quantities of groundwater must be exploited to meet the ever-increasing potable water demand. To this end, an evaluation of the quality groundwater in the area becomes essential. Poor drinking water quality, according to [7] can expose humans to potential health risks. Groundwater generally is considered to be of better quality, compared to surface water in terms of potability. Deterioration in quality, however, may occur over time due to certain natural processes and anthropogenic activities. According to [10], some of the natural processes may include hydrogeochemical processes such as water-rock-interaction, soil-water interaction, cationic exchange reactions, mixing of waters; certain biological processes including selective uptake by vegetation and evapotranspiration. Anthropogenic activities that may result in leachates into groundwater resources may come from agricultural fields that utilize especially, agro-chemicals and manures, municipal and urban wastewater sites, mining activities and mine waste disposal sites, domestic and industrial effluents, etc.

Consequently, this study is aimed at assessing the quality of groundwater resources for drinking purposes and to evaluate the potential risk(s) to human health in the Ahafo-Ano South District of Ghana, which is an artisanal gold mining area. In the authors' view, this will also provide a framework for any future groundwater quality monitoring and evaluation exercise to ensure sustainable utilization of the resources.

2. Materials and methods

2.1 The study area

2.1.1 Location, climate and drainage

The Ahafo-Ano South District (**Figure 1**), is located between longitude 1°45′ and 2°20'W and latitude 6°42′ and 7°10'N with an area of about 1190.7km^2 and an estimated population as at 2010 census to be approximately 121, 659. The district is bordered to the north-east by Tano-North District, north-west by Ahafo-Ano North District; south by Atwima Nwabiagya district and to the east by Offinso-North Municipal Topography is generally undulating with dominant hills being Aya, Kwamisa and Tinte hills, which run from the west towards the northeast. The highest elevation is about 763 meters above mean sea level [11]. According to [12], the area falls within the wet semi-equatorial climatic region of Ghana, which is characterized by the occurrence of two rainy seasons. The major season occurs normally between March and June while the minor season spans September to November. The average annual precipitation is about 1,700 mm - 1850 mm per year while the mean annual temperature is around 30°C with the lowest temperature being about 26.1°C. Relative humidity range is 70–75%. The area lies within the semi-deciduous forest belt with

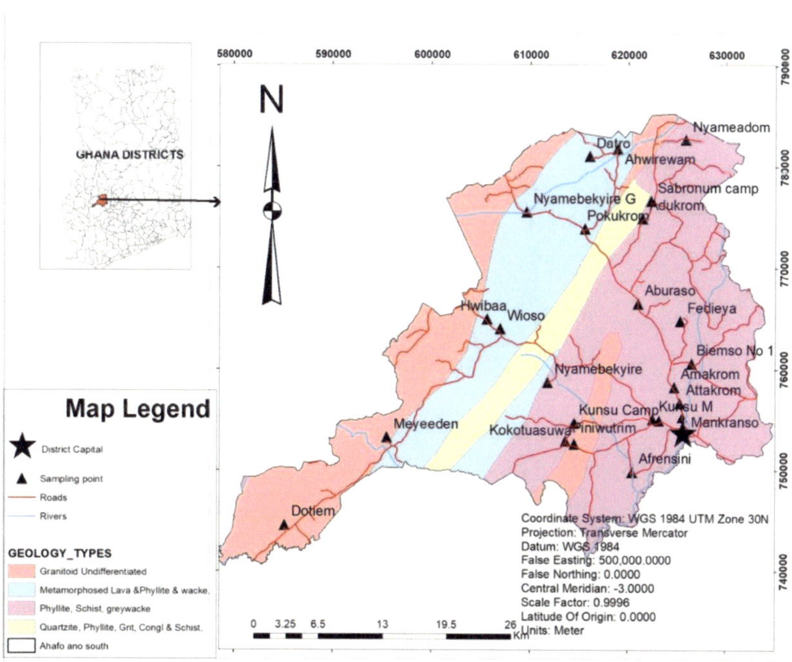

Figure 1.
Geology map of the study area.

vegetation types controlled by precipitation and groundwater. Prominent rivers draining the area include Mankran, Abu and Aboabo rivers and their tributaries.

2.1.2 Geology, soil and natural resources

The area is underlain by the paleo-proterozoic Birimian metasediments subgroup. Common rock include phyllites and schist-intruded with their syngenetic basin-type granitoids [13]. The area lies within the Kumasi-Offin-Dwinyama-Bechemso belt [14], which contains extensive mineral deposits of great commercial significance such as gold, bauxite, manganese deposits as well as extensive granitic outcrop within which commercial quarrying operations occur. Gold-bearing rocks are very widespread and are commonly found and mined communities including Afrensini, Piniwutrim, Amakrom, Aburaso, Ahwirewam, Nyamebekyire, Kunsu, Mankranso, Sabronum, Barniekrom and many other parts of the District. The district is one the most important when it comes to natural resources in Ghana. It large forest reserves, namely, Tinte, Tano, Opuro River, Kwamisa forest reserves, parts of Asufufu Basin and Offin-North forest reserves. The study area is a major food basket in Ghana. About 80% of the land is arable, over 60% of which is used for agriculture. The cultivation of cash crops such as cocoa and cashew is very extensive with food crops such as corn, rice, bananas, cassava, as well as vegetables such as tomatoes, cabbage, lettuce. As briefly highlighted elsewhere, the district is known for extensive illegal artisanal gold mining operations due to the widespread gold deposits. Bauxite deposits are found at Aya Hills and Mpasaaso while manganese deposits can be found at Asirebuo Camp, close to the south of Mpasaaso. There are also clay deposits at Hwibaa, Wioso, Asuadei, Biemso II and Mankranso which have been tested by the Building and Road Research Institute (BRRI) as one of the richest clay deposits in the country [11].

The distribution of soils in the area is as shown in **Figure 2**. Three different soil types are commonly found in the area, namely Ferric Acrisols Rhodic Nitosols and

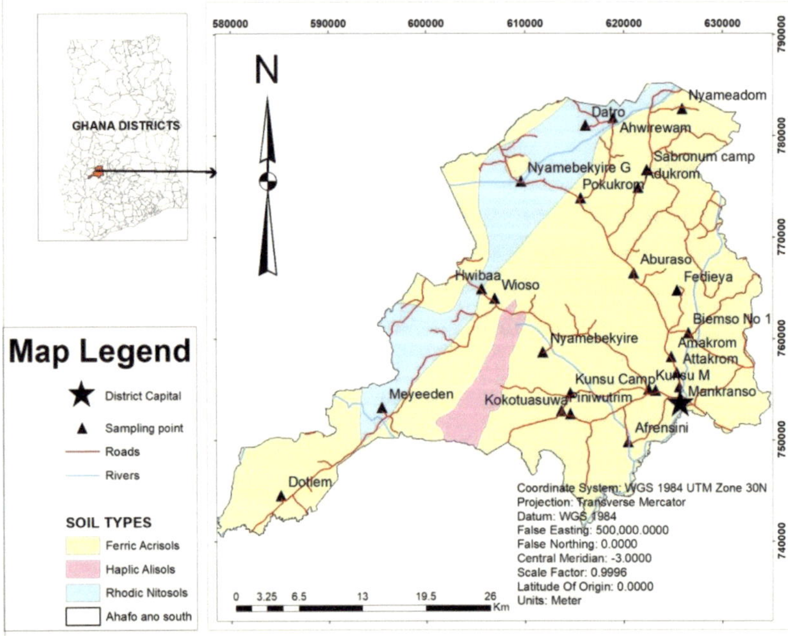

Figure 2.
Soil map of the study area.

Haplic Alisols [15] (**Figure 2**). Ferric acrisols are the dominant soil in the district follow by nitosols and then alisols. Ferric acrisols is a clay-rich soil and is associated with humid, tropical climates and it mostly supports forest areas. Limited cultivation occurs in areas underlain by acrisols due to their low fertility and the toxic amount of aluminum. Nitosols is a deep, red, well-drained soil with a clay content of more than 40% and dominantly from moderate to strong well-develop fine to mouth angular blocky structure [16].

2.2 Methods

2.2.1 Water sampling and analysis

The selections of sampling sites (**Figure 3**) were based on a number of factors such as the ongoing land use activities (farming and illegal artisanal gold mining) and geological settings. Twenty-four (24) well-distributed groundwater point sources (boreholes) functioning hand pumps within the study area according to protocols developed by [17, 18] were sampled in 0.5 litre polythene containers. The containers were conditioned by thoroughly washing with detergent, rinsed several times with acidified water (2%) HNO_3 to prevent contamination. Boreholes were purged for over five (5) minutes so as to obtain fresh samples that were filtered through a 0.45-micron membrane. At each well-site, two (2) samples were obtained for the determination of heavy metals and major ions. To analyze for heavy metals in the laboratory, one sample is filtered and acidified with 2% v/v of HNO_3 to keep ions in solution while unacidified samples were used for major cation and anion analysis. Electrical conductivity (EC), pH and alkalinity were measured at point sampling in the field, using a calibrated WTW field conductivity meter model LFT 91, WTW

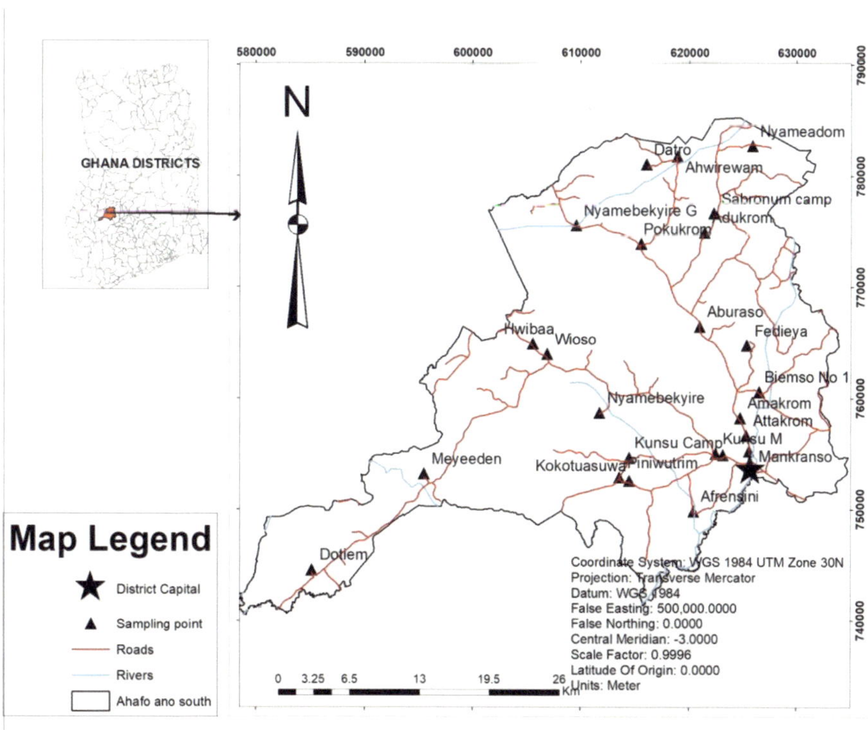

Figure 3.
The study area showing water sampling points.

field pH meter model pH 95 and a HACH digital titrator respectively. A multipurpose electronic DR/890 Colorimeter was used to measure the color, turbidity, total dissolved solids. Sodium and Potassium were analyzed in the laboratory using the flame photometer. Calcium and Magnesium were analyzed using the AA240FS the Fast Sequential Atomic Absorption Spectrometer whilst ICS-90 Ion Chromatograph (DIONEX ICS -90) was used for carrying out chloride (Cl), fluoride (F), nitrate (NO_3) and sulphate (SO_4) analyses. Phosphate was determined by the ascorbic acid method using the ultraviolet spectrophotometer (UV-1201). Five (5) ml of each acidified water sample was measured and 6 ml of nitric acid, 3 ml of HCl and 5 drops of hydrogen peroxide (H_2O_2) were added for acid digestion and placed in a milestone microwave lab station ETHOS 900. The digestate was then assayed for the presence of Zinc (Zn), lead (Pb), Copper (Cu), Chromium (Cr) and Cobalt (Co) using VARIAN AAS240FS Atomic Absorption Spectrum in an acetylene-air flame. Arsenic (As) and Mercury (Hg) were determined using argon-air flame.

2.2.2 Estimation of water quality index (WQI)

To evaluate the general suitability of groundwater for drinking purposes, the water quality index (WQI), a rating and index concept, which was originally proposed by [19] as has been found to be a widely acceptable approach. According to [20], WQI is a rating that reflects the composite influence of different water quality parameters on a picture of the quality of groundwater for most domestic uses. The estimation of WQI requires the utilization of significant parameters that may influence the purpose to which the water is required. Major cations and anions as well as heavy metals that may impose health implications to human health selected are pH, Electrical Conductivity (EC), Sodium, and Calcium, Nitrate, Fluoride, Chloride, Sulphate, Zinc, Lead and Cadmium, and were used to calculate the WQI for this study. The highest weight of five (5) was assigned to lead, nitrate, and fluoride due to their health significance to human health. The water quality index (WQI) for each groundwater source (borehole) is estimated using the following series of relations (1, 2, 3 and 4), and the result was compared to the criteria defined by [20];

$$W_i = \frac{w_i}{\sum_{i=1}^{n} w_i} \tag{1}$$

where, Wi is known as the relative weight; wi is the assigned weight to an influential parameter relative to its impact on the overall quality for drinking purpose and also health implications to humans; $\sum_{i=1}^{n} w_i$ is the summation of assigned weights of all the influential parameters

$$q_i = \frac{C_i}{S_i} \times 100 \tag{2}$$

where, q_i is referred to as the water quality rating; C_i and S_i represent the measured concentration in sampled groundwater and the respective standard [7] of the ith influential parameter. The water quality sub-index for each of the influential parameter (SI_i) is estimated as

$$SI_i = q_i \times W_i \tag{3}$$

where, the symbols have their usual meanings.

$$WQI = \sum_{i=1}^{n} SI_i \tag{4}$$

Thus, the sum of all the water quality sub-indices is resulting in the estimated water quality index for i^{th} influential parameter.

3. Results and discussion

The summarized results of the analyzed groundwater samples are shown in **Table 1**.

3.1 General properties of groundwater

With the exception of pH and TH, all the measured physical and chemical parameters (both cations and anions) of groundwater in the study area fell within their respective acceptable limits for drinking water for humans. The pH of

Parameter	Unit	Min	Max	Mean	[7]
Temperature	°C	27.9	29.700	28.728	**N/A**
pH	pH units	**4.42**	8.540	7.423	**6.5–8.5**
Turbidity	NTU	0.000	2.000	0.926	5
TDS	mg/L	25.000	573.000	175.185	**1000**
Conductivity	uS/cm	48.900	990.000	317.185	**500**
TSS	mg/L	0.000	3.000	0.481	5
Hardness	mg/L	48.000	**348.000**	186.222	**150–300**
Salinity	mg/L	0.000	0.500	0.104	**n.a**
Na^+	mg/L	7.000	98.000	26.519	**200**
Ca^{2+}	mg/L	8.000	59.200	32.000	**200**
Mg^{2+}	mg/L	29.510	83.901	46.094	**150**
K^+	mg/l	2.000	**36.000**	16.778	**50**
HCO_3^-	mg/L	19.520	363.510	137.614	**N/A**
SO_4^{2-}	mg/L	8.000	185.000	33.000	**400**
Cl^-	mg/L	6.000	102.000	27.474	**250**
NO_3^-	mg/L	0.080	0.520	0.143	**50**
PO_4^{3-}	mg/L	0.042	3.573	1.900	**30**
F^-	mg/L	0.680	1.450	1.165	**0.5–1.5**
Pb	Mg/l	0.000	**0.280**	0.048	**0.01**
As	mg/L	0.003	**0.308**	0.042	**0.01**
Fe	mg/L	0.070	**1.050**	0.298	**0.3**
Ni	mg/L	0.000	**0.410**	0.059	**0.02**
Cd	mg/L	0.000	**0.030**	0.059	**0.003**

Table 1.
Summary of water quality analysis results.

twenty-three (23) samples, representing about 92% of groundwater in the area fell within the [7] acceptable range of 6.5–8.5 for potable water while two communities (Sabronum Camp and Amakrom) representing 8% had values within the range of 4–6.5 with respective values of 4.42 and 4.48.

Thus, according to [21], 92% of groundwater was neutral while only 8% were moderately acidic. Hardness varied from 48 to 348 mg/l with a mean of 186.22 mg/l, with the exception of two (2) samples from Adukrom and Ahwirewan representing 8% exceeding the recommended limit of 300 mg/l [7] for drinking water. Water with hardness exceeding 300 mg/l, according to [22] may cause encrustation. Furthermore, [23] classified groundwater based on hardness as soft when hardness is less than 75 mg/l, moderately soft if hardness lies between 76 and 150 mg/l, hard when values fall between 150 to 300 mg/l and very hard groundwater if values are greater than 300 mg/l. In this study, 4%, 32%, 56% and 8% of groundwater in the area could be described as being soft, moderately soft hard and very hard respectively. Thus, groundwater could be generally considered to be predominantly moderately soft to hard water. TDS fell below the [7] limit of 1000 with values ranging from 25 to 573 mg/l with an estimated mean value of 175.185 mg/l. This according to [24], indicate that groundwater in the study area is fresh and young since TDS and conductivity values were generally within acceptable limits.

3.2 Chemical quality of groundwater as drinking water

The potability of sampled groundwaters in the Ahafo-Ano south District was evaluated using influential parameters (IPs). IPs is considered to be the drinking water quality parameters that have significant implications for the human health when ingested through water and food chain or by dermal contact [25]. By utilizing Eq. (1), the estimated relative weights (W_i) are presented in **Table 2**.

Eqs. (2)–(4) above, were utilized to estimate the respective WQI for each sampled groundwater was estimated and presented in **Table 2**. Based on the estimated WQI values, the potability of each sampled groundwater was appropriately classified by comparing to the scheme described in [20]. According to the scheme, groundwaters with WQI being less than 50 are 'Excellent' for drinking by humans, it 'Good water' when WQI range from 50 to 100; 'Poor water' if WQI range from 100 to 200; 'Very poor quality' if WQI range from 200 to 300 and 'Unsuitable as drinking water' when WQI exceeds 300 from **Table 2**,

The status of drinking water quality of sampled groundwater resources in the study area are shown in **Figure 4**.

Illegal artisanal gold mining is most intensive in Afrensini, Piniwutrim Amakrom with water quality described as unsuitable, unsuitable and very poor, respectively. It can be deduced that the anthropogenic activities in these three (3) communities might have rendered the groundwater qualities unsuitable for human consumption. Qualities of groundwater from ten (10) communities with lesser intensive

IP	pH	Ca	Mg	Na	Cl	SO4	Fe	As	Pb	Ni	Cd	$\sum w_i$
S_i	8.5	200	150	200	250	400	0.3	0.01	0.01	0.02	0.003	
w_i	4	1	2.00	2	3	3	3	5	5	3	4	**39**
W_i	0.103	0.026	0.05	0.051	0.077	0.077	0.077	0.128	0.128	0.077	0.103	

Table 2.
Assigned and estimated relative weights.

DOI: http://dx.doi.org/10.5772/intechopen.100242

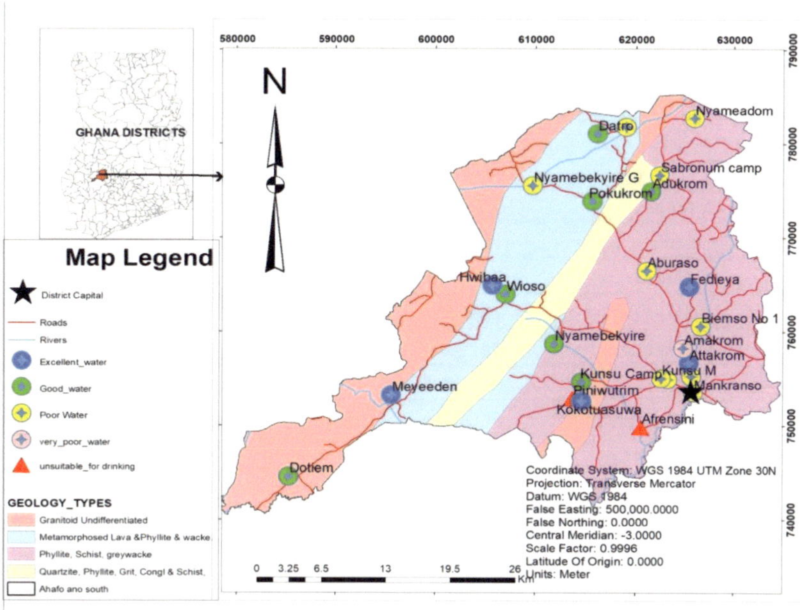

Figure 4.
Spatial distribution of drinking water quality of sampled groundwater.

illegal artisanal gold mining activities were described as poor. Conversely, sampled groundwaters from five (5) and seven (7) communities (representing 20% and 28% respectively) where illegal gold mining operations were non-existent had excellent and good drinking water qualities respectively. Areas with excellent drinking quality include Attakrom, Meyeeden, Kokotuaso, Fedieya and Hwibaa while areas with good groundwater drinking quality include areas include Wioso, Nyamebekyire, Adukrom, Pokukrom, Kunsu camp, Datro and Dotiem (**Table 3**).

3.3 Heavy metals and health implications

Figure 5 presents the spatial distribution of analyzed heavy metals (Pb, As, Cd, Fe and Ni) in the study area. There was an observed elevated level of heavy metals above their respective permissible limits for drinking purposes as defined by [7]. For instance, eight (8) groundwater samples (representing 32% each) had levels of Fe and Cd exceeding their permissible limits of 0.3 mg/l and 0.003 mg/l respectively. The strong correlation between As-Pb (0.73) and As-Ni (0.5) as shown in **Table 4** could indicate a possible similar source. Twenty-one (21) samples out twenty-five had Pb concentrations exceeding [7] permissible limit of 0.01 mg/l while ten (10) and three (3) samples had Ni and As concentrations exceeding their permissible limits of 0.02 mg/l and 0.01 mg/l, respectively.

The observed unsuitable, very poor and poor groundwater drinking quality could therefore be due to the heavy metals such as Pb, Cd, Ni and As, which might have been released into surface waters and soils as a result of improper disposal of mined wastes by the illegal gold miners. Within the Piniwutrim, Afrensi, Amakrom and surrounding villages for instance, virtually, all streams and perennial rivers have been heavily polluted and their courses virtually blocked. In such situations, infiltration of surface waters laden with heavy metals may be enhanced, leading to the leaching of heavy metals infiltrating alongside the surface waters into the groundwater system.

No	Community	N	E	WQI	Classification
1	Mankranso	753517	625904	132.38	Poor water
2	Kunsu Engineer	754916	623278	192.25	Poor water
3	Wioso	763951	607010	66.95	Good water
4	Hwibaa	764903	605689	38.94	Excellent water
5	Nyamebekyire	758695	611902	77.15	Good water
6	Fedieya	764743	625492	47.90	Excellent water
7	Aburaso	766466	621102	134.52	Poor water
8	Sabronum camp	776640	622439	175.79	Poor water
9	Nyameadom	782722	626003	102.15	Poor water
10	Adukrom	774933	621565	67.39	Good water
11	Ahwirewam	781807	619012	112.44	Poor water
12	Nyamebekyire	775544	609697	174.66	Poor water
13	Pokukrom	773880	615706	77.38	Good water
14	Afrensini	749813	620545	363.56	Unsuitable
15	Kokotuasuwa	752580	614664	38.52	Excellent water
16	Piniwutrim	752848	613756	462.69	Unsuitable
17	Kunsu Camp	754657	614639	61.32	Good water
18	Kunsu M	755030	622626	108.26	Poor water
19	Bronikrom Pt 1	755248	625714	158.90	Poor water
20	Meyeeden	753206	595590	39.13	Excellent water
21	Biemso No 1	760573	626656	151.08	Poor water
22	Datro	781059	616186	87.21	Good water
23	Attakrom	756652	625429	47.23	Excellent water
24	Dotiem	744527	585245	83.67	Good water
25	Amakrom	758242	624895	286.82	Very poor water

Table 3.
Estimated WQI and classification.

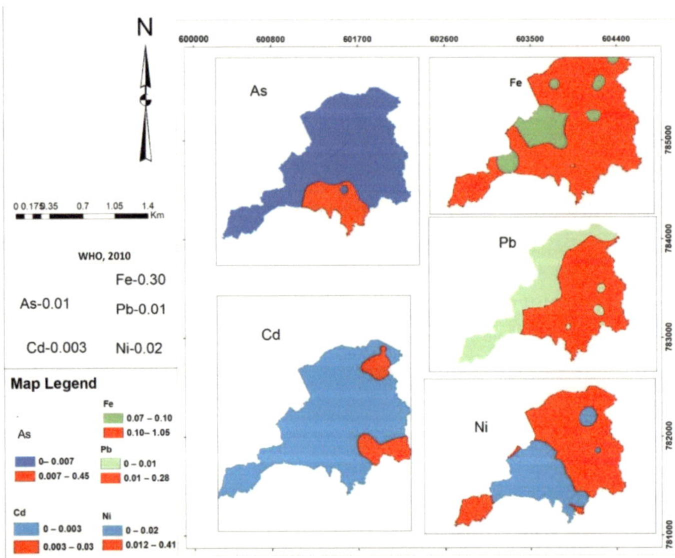

Figure 5.
Distribution of heavy metals in the study area.

	pH	Cond	Temp	TSS	Turb	TDS	Sal	TH	Alk	Fe	Cd	Ni	Pb	Cl	Na	K	SO_4	Ca	HCO_3	CO_3	Mg	As
pH	1																					
Cond	0.37	1																				
Temp	0.14	0.04	1																			
TSS	0.2	0.13	0.16	1																		
Turb	0.38	0.07	−0.1	−0.4	1																	
TDS	0.42	**0.89**	−0.1	0.19	0.15	1																
Sal	0.33	**0.75**	−0.1	0.03	0.13	0.9	1															
TH	0.4	0.35	0.59	0.17	0.13	0.35	0.36	1														
Alk	0.61	0.46	−0.3	−0.1	0.3	0.56	0.48	0.14	1													
Fe	0.09	−0.1	−0.3	0.16	0.41	0.03	0.01	−0.3	0.09	1												
Cd	−0.1	0.26	0.37	−0.3	0.11	0.24	0.33	0.34	0.07	−0.3	1											
Ni	0.19	−0.1	−.01	0.02	0.28	−0.1	0.01	−0.4	0.05	**.494**	−0.2	1										
Pb	0.04	0.21	−0.1	0.12	0.13	0.11	0.12	0.25	0.09	0.22	0.07	−0.3	1									
Cl	0.22	**0.48**	−0.1	0.05	0.15	**0.52**	0.62	0.36	0.21	0.04	0.06	−0.2	0.3	1								
Na	0.34	**0.48**	0.05	0.18	0.29	0.48	0.52	0.37	0.2	0.13	0.08	−0	0.21	**0.91**	1							
K	0.33	0.29	0.24	0.17	0.23	0.34	0.48	0.59	0.31	−0.1	0.17	−0	0.09	0.4	0.5	1						
SO_4	0.42	**0.62**	0.29	0.25	0.16	0.62	0.6	0.41	0.27	0.08	0.17	0.12	−0.1	0.54	**0.64**	0.36	1					
Ca	0.29	0.13	**0.58**	0.27	−0.2	0.07	0.14	0.69	0.2	−0.4	0.16	−0.3	0.17	0.07	0.12	0.51	0.23	1				
HCO_3	**0.6**	0.45	−0.4	−0.2	0.31	0.55	0.48	0.12	1	0.1	0.07	0.06	0.07	0.23	0.22	0.33	0.27	−0.1	1			
CO_3	0.6	0.45	−0.4	−0.2	0.31	0.55	0.48	0.12	1	0.1	0.07	0.06	0.07	0.23	0.22	0.33	0.27	−0.1	1.0	1		
Mg	0.4	0.34	**0.59**	0.16	0.14	0.34	0.36	1	0.14	−0.3	0.34	−0.4	0.24	0.38	0.39	0.61	0.42	**0.69**	0.12	0.12	1	
As	−0.1	0.19	0.08	0.25	−0.1	0.22	0.19	**0.45**	0.01	0.14	0.18	**0.46**	**0.73**	0.4	0.35	0.23	0.03	0.39	−0.1	−0.1	0.45	1

Table 4.
Spearman correlation matrix of analyzed water quality parameters.

3.4 Health effect due to elevated heavy metals in groundwater

The distribution of Pb in groundwater appears to be quite widespread. Twenty-one (21) representing about 84% of sampled groundwater elevated Pb levels exceeding [7] limit of 0.01 mg/l for potable water in the study area. This presents a potential health hazard to the inhabitants. According to [26], this is because, Pb normally accumulates in human body and builds up over a long period of exposure with no known documented usefulness to human health regardless of the pathway. But instead, exposure to different levels of Pb may predispose children, pregnant women and adults in general to serious health consequences [7]. According to [27], low levels of Pb (Pb $\leq$ 5 μg/dl) may lead to hearing impairment, malfunctioning of blood cells in children, low IQ and damage to central and peripheral nervous systems. In pregnant women, it may lead to reduced growth of foetus and premature birth. This may put the lives of both the unborn baby and mother at risk. Pb may cause delayed puberty and reduction in IQ in children while there could be incidences of increased blood pressure when humans are exposed to Pb concentrations between 5 and 10 μg/dl. According to [26], exposure to Pb levels exceeding 15 μg/dl may predispose human to nervous disorders, reduced kidney function, low sperm count in men, and delayed conception in women. In this study, measured Pb concentrations fall within 'low Pb level' indicating that ingestion of groundwater by human may predispose especially children and pregnant women to potential health risks such as low IQ, damaged central and peripheral nervous system, impaired hearing and functioning of blood.

The concentration of Ni in sampled groundwaters ranged from 0 to 0.41 mg/l with a mean of 0.059 mg/l. Ten (10) out of the twenty-five (25) groundwater samples, representing about 40% had levels of Ni exceeding [7] permissible limit of 0.02 mg/l. According to [28], the commonest sources of Ni in mining areas may include effluent water generated from mining and smelting operations, runoff from tailing piles, or from utility water used for mine operations. Nickel has an extensive range of carcinogenic mechanisms which include regulation of transcription factors, controlled expression of certain genes and generation of free radicals. Nickel has been shown to be implicated in regulating the expression of specific long non-coding ribonucleic acids (RNA). It has also been demonstrated that nickel can generate free radicals, which contribute to carcinogenic processes. Common adverse health effects when Ni is ingested at high concentrations on humans include chronic bronchitis, lungs and nasal sinus cancer and reduced lung function. The exposure to nickel at high concentrations can be alarming as it causes chronic bronchitis, lungs and nasal sinus cancer and reduced lung function [29].

Cd concentration in the area varied from 0 to 0.03 mg/l with average value of 0.059 mg/l. In the current study, about 32% of groundwater had Cd levels above recommended limits for drinking water. High levels can be linked to the weathering and subsequent dissolution of chalcopyrite and pyrite ores in the area [30]. These minerals are common pathfinder minerals of gold within the meta-volcanics of Ghana [14]. The ingestion of Cd at levels exceeding the [7] permissible limits for drinking water can cause stomach irritation resulting to vomiting and diarrhea, and also lead to degenerate bone disease such as osteoporosis (skeletal damage). Short-term exposure to inhalation may lead to severe damages to the lungs and respiratory irritation while long-term exposure may lead to deposition in bones and lungs. Cd is highly toxic to the kidney and when accumulated in the proximal tubular cells at higher concentrations may predispose humans to renal dysfunction and kidney disease. Cd is also classified as group 1 carcinogens for humans by the International Agency for Research on Cancer [31].

Values of As in this study varied from 0.003 mg/l to 0.308 mg/l with an average concentration of 0.042 mg/l. Three samples out of twenty-five (25) sampled groundwater, representing about 12% had As values exceeding the recommended limit for potable water. Higher levels of As in groundwater may be due to anthropogenic activities such as mining and processing of gold ores, and naturally from water-rock interactions or soil-water interactions [7]. According to [32], As is commonly associated with ores containing metals, such as copper and lead. The strong positive correlation (0.73) between As and Pb concentration in this study supports this assertion and may indicate a common source. Exposure of As at lower levels can cause nausea and vomiting, reduced production of erythrocytes and leukocytes, abnormal heartbeat, pricking sensation in hands and legs, and damage to blood vessels. Long-term exposure may result in the occurrence of skin lesions, internal cancers, neurological problems, pulmonary disease, peripheral vascular disease, hypertension and cardiovascular diseases, and diabetes mellitus [33].

4. Conclusion

The activities of illegal artisanal gold mining have led to widespread degradation in the quality of groundwater in the study area as a result of indiscriminate dumping of mined wastes. Due to the potential interaction between contaminated surface waters and groundwater resources, it is anticipated that surficial waters with deteriorated quality arising from the contamination from the illegal small scale mining activities in the area may have compromised the otherwise generally better quality groundwaters in the area, which serve as the main source of potable water. This study has evaluated the quality of groundwaters to ascertain the potability and potential health risks associated with the consumption of groundwater resources in the Ahafo-Ano South District of Ghana. It was observed that areas with intensive illegal artisanal gold mining operations had much higher elevated levels of heavy metals such as Pb, As, Cd and Ni in sampled groundwaters and vice-versa. It is expected that the findings of this study will present a framework upon which, all stakeholders involved in the management of water resources and water supply for the inhabitants will develop an effective, efficient as well as preventive mechanisms to minimize groundwater quality deterioration, minimize potential health risks and to enhance the socio-economic development of the people.

Author details

Maxwell Anim-Gyampo[1], Belinda Seyram Berdie[2], Enoch Bessah[3] and Sampson Kwaku Agodzo[3*]

1 Department of Geological Engineering, School of Engineering, University for Development Studies, Nyankpala Campus, Tamale, Ghana

2 Earth Science Department, Faculty of Earth and Environmental Sciences (FEES), C.K. Tedam University of Technology and Applied Sciences, Navrongo, Ghana

3 Department of Agricultural and Biosystems Engineering, Faculty of Mechanical and Chemical Engineering, College of Engineering, KNUST, Kumasi, Ghana

*Address all correspondence to: skagodzo7@usa.net

References

[1] World Bank, 2013. Artisanal and Small Scale Mining. http://www. worldbank. org/en/topic/ extractiveindustries/brief/artisanal-and-small-scale-mining. (Accessed 15th November, 2020; Time: 14:43)

[2] Amankwah, R. K. and Anim-Sackey, C., (2003). Strategies for sustainable development of the small-scale gold and diamond mining industry of Ghana. Resources Policy. 29 , pp31-138. doi:10.1016/j.resourpol.2004.07.002.

[3] Adu-Gyamfi, E., (2014). The effect of illegal mining on school attendance and academic performance of junior high school students in Upper Denkyira-West District of Ghana. Journal of Education and Human Development, 3, 5523-5545.

[4] Hilson, G., Hilson, A. and Adu-Darko, E., (2014). Chinese participation Ghana's informal gold mining economy.

[5] Kabata-Pendias, A. Trace Elements in Soil and Plants, 4th ed.; Taylor & Francis: Boca Raton, FL, USA, 2011.

[6] Ghana Minerals Commission, (1995). Project Implementation Plan: Mining Sector Development and Environment Project.

[7] World Health Organization, (2010). Guidelines for drinking water quality WHO, Vol 1, Recommendation, Geneva.

[8] Eduful, M., AlSharif, K., Acheampong, M. and Nkhoma, P., (2020). Management of catchment for the protection of source water in the Densu River basin, Ghana: implications for rural communities. International Journal of River Basin Management. http://doi.org/10.1080/15715124.2020.1750420.

[9] Anim-Gyampo, M., Anornu, G. K., Appiah-Adjei, E. K. and Agodzo, S. K., (2018). Hydrogeochemical evolution and quality assessment of groundwater within the Atankwidi basin: the case of northeastern Ghana. Arabian Journal of Geosciences 11: 439 https://doi. org/10.1007/s12517-018-3753-6.

[10] Appelo, C. A. J., and Postma, D. P., (2005). Geochemistry, groundwater and pollution, 2nd edn. Balkema, Rotterdam, 647p.

[11] Ghana Statistical Service, (2012). 2010 Population and Housing Census. Summary Report of Final Results. Ghana statistical Service, Accra, Ghana.

[12] Dickson, K. B. and Benneh, G., (1998). A New Geography of Ghana. Longman Group UK Limited, Longman House Burnt Mill, Harlow, pp. 27-52.

[13] Griffis, J. R., Barning, K., Agezo, F. L. and Akosah, K. F., (2002). Gold deposits of Ghana. Graphic Evolution in Barrie, Ontario, 362p.

[14] Kesse, G. O., (1985). The Mineral and Rock Resources of Ghana. Balkema Publishing, Rotterdam, 600p.

[15] Ghana Shape Files, (2009).

[16] World Health Organization/Food and Agriculture Organization, (2007). Joint FAO/WHO Food Standard Programme Codex Alimentarius Commission 13th Session. Report of the Thirty eight Session of the Codex Committee on Food Hygiene, Houston, United States of America, ALINORM 07/30/13.

[17] Claasen, H. C., (1982). Guidelines and techniques for obtaining water samples that accurately represent the water quality of an aquifer. US Geological Survey, open-file report 82-1024, 49p.

[18] Barcelona, M., Gibb, J. B., Helfrich, J. A. and Garske, E. E., (1985) Practical

guide for groundwater sampling. Illinois State Water Survey ISWS, Contract Report 374.

[19] Horton, R. K., (1965). An index number system for rating water quality. J Water Pollut Control Fed 37(3): 300-306

[20] Sahu, P. and Sikdar, P. K., (2008). Hydrogeological framework of the aquifer in and around East Kolkota Wetlands, West Bengal, India. Environmental Geology, 55: 823-835. http://doi 10.1007.s00254-007-1034-x.

[21] Hounslow, A. W., (1995). Water quality data: Analysis and Interpretation. Florida, U.S.: CRC Press Inc.

[22] McGowan, W., (2000). Water processing, residential, commercial, lightindustrial, 3rd edn. In: Harrison JF (Ed) Water quality association. lisle, Illinois pp 309.

[23] Saravanakumar, K. and Kumar, R., (2011). Analysis of water quality parameters of groundwater near Ambattur industrial area, Tamil Nadu, India. Indian Journal of Science and Technology, 4(5):560-562.

[24] Freeze, R. A. and Cherry, J. A., (1979). Groundwater. Prentice-Hall, Englewood, 624p.

[25] Yidana, S and Yidana, A., (2010). Assessing groundwater quality using water quality index and multivariate statistical analysis – the Voltaian basin, Ghana. Journal of Environmental Earth Science. 59 (7), pp. 1461-1473. http://doi.org/10.1007/s12665-009-0132-3.

[26] Center for Disease Control and Prevention, (2012). What do parents need to know to protect their children? Available: http://www.cdc.gov/eeh/lead/ACCLPP [accessed 16th November, 2020, Time: 17:42].

[27] United States Environmental Protection Agency, (2017). Risk assessment guidance for superfund, Vol. 1, human health evaluation manual (part A), Report EPA/540/1089/002, Washington, DC

[28] Mann, H., Fyfe, W. S. and Kerrich, R., (1989). Retardation of toxic heavy metal dispersion from nickel–copper mine tailings, Sudbury district, Ontario: role of acidophilic microorganisms. I. Biological pathway of metal retardation. Biorecovery 1:155-172

[29] Agency for Toxic Substances and Disease Registry, (ATSDR) (2005) Toxicological profile for Nickel U.S. Department of Health and Human Services, Public Health Service, Division of Toxicology 1600, Atlanta, GA 30333.

[30] Autier, V. and White, D., (2004). Examination of cadmium sorption characteristics for aboreal soil near Fairbanks, Alaska. J Hazard Mater 106B:149-155.

[31] Davies, B. E., Bowman, C., Davies, T. C. and Sellinus, O., (2005). Medical geology: perspectives and prospects. Essential Medical Geology, Elsevier Inc., pp. 1-14.

[32] Agency for Toxic Substances and Disease Registry, (2007). U.S. Department of Health and Human Services, Public Health Service, Division of Toxicology 1600, Atlanta, GA 30333

[33] Smith, A. H., Lingas, E. O. and Rahman, M., (2000). Contamination of drinking water by arsenic in Bangladesh: a public health emergency. Bull World Health Organ 78(9):1093-1103.

Chapter 7

Mobility of Heavy Metals in Aquatic Environments Impacted by Ancient Mining-Waste

Marilú Barrera Olivarez, Mario Alfonso Murillo Tovar, Josefina Vergara Sánchez, María Luisa García Betancourt, Francisco Martín Romero, América María Ramírez Arteaga, Gabriella Eleonora Moeller Chávez and Hugo Albeiro Saldarriaga Noreña

Abstract

The mobility of heavy metals in aquatic environments, impacted by discharges from mining waste, is one of the major processes causing metal pollution mainly by arsenic (As), cadmium (Cd), lead (Pb), zinc (Zn) and iron (Fe), which could be risky for biota and human health. The heavy metals contained in mining waste constituted by large amounts of sulfides can reach the aquatic compartments by acid mine drainage and runoff and eventually become deposited in sediments and associated with colloidal material, being this one of the main reservoirs and ways of transport. However, the mobility of heavy metal is influenced by their specific chemical properties and undergo several physicochemical phenomena as sorption, oxidation–reduction, hydrolysis and this can be influenced by water flow, the size and composition of geological material. Hence, this work aims to review the processes and mechanism involved in the fate and transport of heavy metals from mining-waste to aquatic compartments and the methods used for identification of the specific chemical species associated with their mobility and ecological risk.

Keywords: mobility, heavy metals, mining-waste, hardpads, acid mine drainage, sediments

1. Introduction

Heavy metals in aquatic environments have mainly a natural origin due to the geological parent material (lithogenic), they can be incorporated into different materials as silicates, carbonates, oxides, hydroxides, and sulfides structures and as a native element. They also could result from anthropogenic sources including deposition of particles (<30 μm in diameter) and precipitation containing heavy metals, fertilizers application, the use of agrochemicals, spilled of wastewater and mining waste [1, 2]. Among anthropogenic sources, ancient mine residues, has a high impact and pose a threat to the environment and health as a consequence of the rustic extraction method and their high content of heavy metals such as arsenic (As), lead (Pb), cadmium (Cd), mercury (Hg), copper (Cu), zinc (Zn) and Iron (Fe), which can cause high damage to aquatic biota and human health [3, 4].

Metals are partitioned among the various aquatic environmental compartments (water, suspended solids, sediments and biota) and can occur in dissolved, particulates or complex form. The metals and metalloids can reach the aquatic environments from mining waste as metallic ions and complexes in dissolved solid formed either by weathering, erosion and run off processes. Once a metal reaches an aquatic reservoir, it does not suffer any degradation, rather they are accumulated in sediments and depending on their chemical form it can increase or reduce their toxicity, bioavailability, and solubility [5–7].

Sediments are considered a main sink and means of transport of organic and inorganic pollutants in aquatic environments. It has been found that they have a great capacity to adsorb metals and metalloids present in the aqueous phase and reduce their mobility in the aquatic environment [3, 8, 9]. Among different particles size that constitute sediments, metals are mainly associated with the smallest particles of colloidal size ranging from 0.001 to 0.1 μm in diameter due to their largest surface area and most likely as a consequence of occurrence of ionic exchange sites linked to several chemical species such as humus Fe, Al and Mn, oxyhydroxides, aluminosilicates and some moderately soluble salts such as calcium carbonate ($CaCO_3$).

The mobility of trace elements, from the vadose zone to the aquifers and in rivers and tributaries, also influenced by sorption, oxidation–reduction, hydrolysis, and complexation and chelation processes, determining the transport of highly toxic metals and metalloids in aquatic environments [1]. Adsorption is likely the most important process that determines the mobility of traces metals in aquatic environments, since it supports ions at the interface between the solid and the aqueous phase, the clay and humus material with a negative charge on its surface adsorbs cations, while oxyhydroxides with varying charges on their surface can adsorb cations and anions, respectively.

Hence, this work aims to review the processes and mechanism involved in the dynamics (fate and transport) of heavy metals from mining-waste to aquatic compartments and the methods used for identification of chemical species associated with their mobility and ecological risk.

2. Processes involving in mobility of heavy metals in natural waters

The mobility of metals and metalloids in aquatic compartments is a very complex phenomenon as it involves a great variety of physical, chemical and biological processes mainly determined by pH values, precipitation and dissolution of secondary minerals, sorption–desorption reactions, hydrolysis processes, by oxidation–reduction processes and co-adsorption processes, for instance: graphene oxide a very soluble chemical specie, it is commonly found in aquatic environments, adsorbed on inorganics contaminants such as the hematite and goethite, which co-adsorbed graphene oxide about 92% at pH 3-5 [10].

The degree sorption/desorption of metals depend on time of contact between sorbate and sorbent and oxide aging due to the weathering. The adsorption and precipitation can be simultaneous but can dominate a mechanism due to reaction conditions and metal involved. When the precipitate consists of species derived from both aqueous solution and dissolution of the mineral, it is referred to as a coprecipitate [7].

Overall, Zn, Cd, Cu and Al cations have high concentrations in acidic water and undergone high mobility, while oxyanions such as SO_4^{2-}, AsO_4^{3-}, MoO_4^{2-}, CrO_4^{2-} increase their mobility when water is neutral or alkaline [4, 11].

2.1 Cation exchange adsorption

Non-specific or cation exchange adsorption, also known as physical adsorption, is an electrostatic phenomenon that occur on the surface of clays (kaolinite, illite, montmorillonite, vermiculite, smectite and chlorite) and it is caused by the weathering of olivine, augite, pyroxene, mica and feldspar. Structure of clays are characterized by thick microcrystalline sheets composed of tetrahedral layers of silica and octahedral layers of aluminum in 1: 1 or 2:1 proportion [1].

Cation exchange is carried out by less selective outer-sphere clusters. The cations are bound to the surface of the negatively charged clay through weak covalent bonding independent of the aqueous pH value. This type of phenomenon is reversible in nature, and occurs very quickly, typically controlled by diffusion and electrostatic reactions, while the smallest ions of the aqueous phase are exchanged for larger ions on the surface of clay, for example, Mg^{2+} exchange Al^{3+} and Al^{3+} exchange Si^{4+}. The number of cations reversibly adsorbed per unit weight of adsorbent (e. g. clay) is called cation exchange capacity (CEC) [2].

2.2 Specific adsorption

The specific adsorption is also known as chemisorption or surface complexation. It is mainly carried out in oxides of Fe, Al, and Mn. The ions, either cations or anions, are highly bonded on the surface of the oxides by covalent bonding with oxygen atom or OH groups. The Fe or Mn oxide are electrically charged by the adsorption or release of H^+ ions, from the oxygen atoms at the interface between the mineral and the solution. Because oxides are amphoteric chemical compounds, they have negative and positive charges on their surface, and the net charge is largely symmetric about at zero point, at a characteristic pH value. The pzc, stand for zero point of charge varies between several oxides compound; Fe oxides have a pzc between pH 7.0 and 8.5 which implies a positive charge on their surface. While the pzc of Mn oxides varies between pH 1.5 and 4.6, which indicates that they have a net negative charge on their surface [1]. It is carried out on the surface of Fe and Mn oxides with variable charges and complexation with MO functional groups, by weak electrostatics charges with pH-dependent bonding.

2.3 Hydrolysis

In pure water at 25°C, $[H_3O^+] = [OH^-] = 1.0 \times 10^{-7}$ M and pH = 7.0 (neutral pH); when dissolving NaCl in water there is no an appreciable hydrolysis reaction and the pH of the solution remains at a value of 7.0. However, when ammonium chloride (NH_4Cl) is added to the water, pH drops below 7.0, it means that $[H_3O^+] > [OH^-]$, and when sodium acetate ($NaC_2H_3O_2$) is dissolved in water, pH increases above 7.0, it means that $[OH^-] > [H_3O^+]$. In general, salts containing an ion of an alkali or alkaline earth metal (except Be^{2+}) do not significantly hydrolyze; thus, when a substance is added to water and dissociated, one of its ions causes a change in the pH of water (pH $\neq 7$), it is at that moment when we speak of a hydrolysis reaction. In fact, all positive ions react with water to produce an acidic solution [12, 13].

$$4Fe^{2+} + O_2 + 10H_2O \rightarrow Fe(OH)_3 \quad (1)$$

$$Fe^{3+} + 3H_2O \leftrightarrow Fe(OH)_{3\ (s)} + 3H^{+} \quad (2)$$

In deposits of tailings containing high levels of sulfides such as pyrite and pyrrhotite, during complex oxidation process, Fe^{2+} dissolves and when reacting with water, carries out a hydrolysis reaction, increasing the acidity of the medium by decreasing the pH value [Eq. (1)] [8, 9]. Likewise, when Fe^{2+} is oxidized to Fe^{3+} and pH values >5, a hydrolysis reaction occurs, releasing H^{+} and ferric hydroxide (reaction 2) that generates an oxyhydroxide [4]. However, the increase in acidity in a mining waste, even though it reaches the aquatic environment, does not necessarily implies that the water has low pH values since the presence of carbonate minerals can neutralize the acidity.

2.4 Oxidation-reduction

Sulfide minerals in mining waste pose several relative resistances to weathering (**Table 1**) [9]. However, due to physicochemical processes, the weathering process presents alterations that influence the composition of water that can drain or percolate in the mining waste.

The oxidation of sulfur minerals is due frequently to their deposition in aerated places where sulfide minerals are thermodynamically unstable. Therefore, pyrite and pyrrhotite, which generally dominate sulfide deposits are the first sulfides to undergo oxidation [Eqs. (3) and (4)] due to their chemical composition. During oxidation process, one mole of pyrite and pyrrhotite with atmospheric oxygen, is the main electron acceptor, and water form Fe^{2+}, 4 and 2 moles of acid, respectively [9]. However, sphalerite, chalcopyrite, galena and arsenopyrite also contribute to the generation of acid drainage [4, 14].

$$FeS_2 + \frac{15}{4}O_2 + \frac{7}{2}H_2O \rightarrow 2SO_4^{2-} + \mathrm{Fe(OH)_3} + 4H^{+} \quad (3)$$

$$Fe_{1-x}\,S + \left(2 - \frac{x}{2}\right)O_2 + xH_2O \rightarrow SO_4^{2-} + (1-x)Fe^{2+} + 2xH^{+} \quad (4)$$

Mineral	Formula	Resistances
Pyrrhotite	$Fe_{(1-x)}S$	Low
Galena	PbS	↓
Sphalerite	(Zn, Fe)S	↓
Bornite	Cu_5FeS_4	↓
Pentlandite	$(Fe,Ni)_9S_8$	↓
Arsenopyrite	FeAsS	↓
Marcasite	FeS_2	↓
Pyrite	FeS_2	High
Chalcopyrite	$CuFeS_2$	
Magnetite	Fe_3O_4	
Molybdenite	MoS_2	

Table 1.
Relative resistances of sulfide minerals and magnetite in oxidized tailings taken from [9].

Heat is released when oxidation reaction of metal sulfides occurs. When pyrite is oxidized and form acid, this reaction releases 1440 $KJ.mol^{-1}$, heat that is hardly released by an oxidation reaction [15]. The heat generated by that exothermic reaction of pyrite, as long as there is high permeability in the tailings cover, it transports oxygen by convection, increasing oxidation rate of sulfides. However, when a reservoir is saturated with water, advective transport is reduced and the main oxygen transport mechanism is diffusion [14, 16]. In addition, oxygen can be supplied by wind in lateral parts of deposit, where oxygen can migrate upward or reach basal regions by advection, convection and diffusion, [17]. Therefore, low concentrations of oxygen are due to its high consumption in oxidation of sulfides minerals and to a limited supply due to the low permeability that governs waste materials.

When the cover of the tailings dam is depleted in oxygen and the air supply is insufficient, high oxidation of sulfides causes water to be characterized by pH <3, high concentrations of sulfates and metals, for instance Zn, Fe, Pb. Under these scenarios, Fe^{3+} ion remains in solution and becomes the dominant oxidant of metal sulfides, generating acidity in the environment, Eqs. (5) and (6) show generation of acid from oxidation of pyrite and pyrrhotite [9].

$$FeS_2 + 14Fe^{3+} + 8H_2O \rightarrow 15Fe^{2+} + 2SO_4^{2-} + 16H^+ \tag{5}$$

$$Fe_{(1-x)}S + (8-2x)Fe^{3+}4H_2O \rightarrow (9-3x)Fe^{2+} + SO_4^{2-} + 8H^+ \tag{6}$$

2.5 Geological factors influencing mobility of heavy metals

Geological formations are volumes of rocks confined to a certain space, they are formed by different types of rocks (igneous, sedimentary, and metamorphic) of different nature, they can enclose mineral deposits of great economic value. Metallic mineral deposits are generally enriched in sulfides such as: pyrite (FeS_2), pyrrhotite ($Fe_{(1-x)}$ S), sphalerite (ZnS), chalcopyrite ($FeCuS_2$), arsenopyrite (FeAsS), galena (PbS) and cubanite ($CuFe_2S_3$). From the metallurgical method, the recovery of high economic value metals such as Pb, Zn and Cu is carried out, releasing to the environment waste ground material commonly called "tailings" enriched in gangue minerals that do not represent any economic interest for its exploitation as: pyrite (FeS_2), arsenopyrite (FeAsS), pyrrhotite ($Fe_{(1-x)}$ S), calcite ($CaCO_3$), quartz (SiO_2) and K feldspars ($AlSi_3O_8$) [3, 18].

Once the deposition of tailings is concluded, either if oxygen and water flow through pores or if there is ferric ion, oxidation of sulfides occur, generating acidity through the release of protons H^+ (Eqs. (7)–(10)]. In deposit of tailings, As is associated with pyrite, Zn and Cd with sphalerite, and they are released when these minerals are dissolved. Despite that galena has a reactivity similar sphalerite, it does dissolve very slowly because secondary mineral of anglesite ($PbSO_4$) generally precipitates on its edges, which avoid its dissolution even in highly oxidized environments removed from sphalerite. Cu in acidic environments is released from chalcopyrite, while Co and Ni are generally derived from the oxidation of pyrite and pyrrhotite [3, 9].

$$FeS_{2\,(s)} + 3.5O_{2\,(ac)} + H_2O_{(l)} \rightarrow Fe^{2+}_{(ac)} + 2SO^{2-}_{4\,(ac)} + 2H^+_{(ac)} \tag{7}$$

$$ZnS_{(s)} + 2O_{2(ac)} \rightarrow Zn^{2+}_{(ac)} + SO^{2-}_{4(ac)} \tag{8}$$

$$PbS_{(s)} + 2O_{2(ac)} \rightarrow Pb^{2+}_{(ac)} + SO^{2-}_{4(ac)} \tag{9}$$

$$FeS_2 + 14Fe^{3+}_{(ac)} + 8H_2O_{(l)} \rightarrow 15Fe^{2+}_{(ac)} + 2SO^{2-}_{4\,(ac)} + 16H^+_{(ac)} \tag{10}$$

If the oxidation of sulfides remains, the acidity would increase at pH <4, generating acid mine drainage (AMD). However, if the mineralogy of the encasing rock has enough carbonate as calcite ($CaCO_3$), hydroxide and silicate, when dissolution occurs, acid is consumed and neutralization of the AMD is completed [Eq. (11)], originating secondary gypsum precipitates and other metal sulfates [9, 11].

$$CaCO_{3\,(s)} + H_2SO_{4\,(ac)} + H_2O_{(l)} \rightarrow CaSO_4 \cdot 2H_2O_{(s)} + CO_{2\,(g)} \tag{11}$$

However, when acidity-consuming minerals are insufficient, neutralization is not achieved, so continuous generation of AMD dissolves mineral phases, causing supersaturation of ions with high electrical conductivity and generating the precipitation of secondary minerals such as oxyhydroxides of Fe^{3+} (goethite), hydroxysulfates (jarosite, scorodite and beudantite). Those highly oxidized tailings, are rich in sulfides; the precipitation of these minerals' forms cemented layers known as hardpads, of low porosity and high density that serve as hydraulic barriers and temporary sinks of metals and metalloids, where the precipitation of beudantite and scorodite limits the mobility of As and Pb, and Fe oxyhydroxides adsorb Cd, Pb, Cr, Zn (despite their high mobility) and As where substitution process in jarosite is common [3, 6, 8]. Although Fe oxyhydroxides have a high adsorption capacity for highly toxic metals, it decreases as increasing crystallinity, the larger grain size the lower surface area. Although high concentrations of metals and metalloids reach the aquatic environment by runoff, their mobility will depend on the nature of the sediments and minerals that predominate in the aquatic environment [9].

3. Chemical fractionation of heavy metals from sediments

In order to obtain information about mobility and potential toxicity of heavy metals or their potentially dangerous effects on the environment, it is necessary to implement a methodology that determines speciation or fractionation of those metals in sediments [19]. Different schemes of chemical fractionation [20] have

Fraction	Reagent or solution	Conditions	Geochemical fraction
Water soluble elements and adsorbed by electrostatic force (F1)	16 mL 1 M of $MgCl_2$	Shakes for 1 hour at room temperature	Salts soluble in water and adsorbed by electrostatic force
Elements associated with carbonates (F2)	16 mL of NaOAc 1 M, adjusted to pH 5 with HOAc	Shakes for 2 hours at room temperature	Associated with secondary carbonates
Elements associated with oxihidroxides (F3)	40 mL of $NH_2OH.HCl$ 0.04 M in HOAc al 25% v/v	90°C with occasional shakes for 4 hours	Oxihidroxidos of primary and secoundary
Elements associated with organic matter and sulfides (F4)	6 mL of HNO_3 0.02 M and 10 de H_2O_2 30% (pH 2 with HNO_3)	85°C for 2 hours	Organic matter and sulfides
Residual (F5)	20 mL of HF and 4 mL of $HClO_4$, 20 of HF and 2 of $HClO_4$, HCl 12 N	The mixture is heated to dryness	Primary minerals such as silicates and quartz

Table 2.
Sequential chemical extraction methodology from Tessier 1979.

been used to identify and quantify concentrations of heavy metals and metalloids present in different fractions. However, the sequential extraction that suggests Tessier 1979, is one of the most used. Each fraction or extract corresponds to the metals associated with water or reagent (acid), Tessier method suggested five chemical fractions: exchangeable fraction (F1), fraction of carbonates (F2), reducible fraction or oxyhydroxides of Fe and Mn (F3), Oxidizable fraction or organic matter and sulfides (F4) and residual fraction (F5) that are extracted with different reagents, under different physicochemical conditions (**Table 2**) [20–22].

4. Discussion

Metals are transported from tailings deposit to the stormwater system through runoff in the soluble or particulate form. Depending on physicochemical conditions of aqueous medium and nature of metal, these can be adsorbed to the colloidal material, precipitate, solubilize or to be complexed. When sequential chemical extraction is applied to sediments impacted by old mining wastes with high concentrations of minerals such as sulfides and significant amounts of carbonates and silicates, it has been determined that Zn, although it has been found in fractions F2 and F3, tends to be more abundant in F1, that is, it has high mobility and availability in the aquatic environment, Cu has more affinity for F3 and F4 fraction [3, 9, 19] and As is generally found adsorbed to the oxyhydroxide fraction [23]. Although in stormwater systems the highest concentrations of heavy metals are found in sediments, rather than in the soluble fraction, these can be available mainly due to changes in pH and cause great effects on aquatic organisms and on the health of the man [24]. On the other head, the fractionation is also a useful tool to determine whether pollutants are of natural origin or anthropogenic. Heavy metals of origin anthropogenic are present mainly in the first fractions, while that the origin lithogenic are in the residual fraction [19].

The old mining residues (tailings), that were deposited near the rivers; they are characterized by having large amounts of sulfides (mainly pyrite and pyrrhotite) with high concentrations of potentially toxic elements (PTEs) [25]. When the sulfides in the tailings are exposed to weathering by the presence of water, atmospheric and dissolved oxygen, oxidation process takes place [26]. The oxidation process, as undergo for pyrite is very complex, when the oxidation reaction is carried out, heat is released, which, by advection, can significantly improve transport of gas in the waste pile, increasing oxidation rate of sulfides [16]. Furthermore, oxygen can also enter lateral parts of reservoir upward and into basal regions. Nevertheless, in the rainy season, tailings deposits can become saturated with water, so diffusion is the main oxygen transport mechanism [14].

In oxidation of pyrite, Fe^{2+} dissolves and reacts with water through hydrolysis process, it generates acidity [9]. On the other hand, when the mining waste cover has insufficient atmospheric oxygen, a high oxidation of sulfides causes water to have a pH <3, then under these acidic conditions, Fe^{3+} can remain in solution and become a dominant oxidant for the oxidation of pyrite [4]. The oxidation of pyrite, causes the dissolution of sulfides such as: sphalerite, pyrrhotite, arsenopyrite and chalcopyrite, although galena has reactivity similar to sphalerite, it does not dissolve easily since an anglesite ($PbSO_4$) edge is formed in galena that is almost insoluble in acidic environments.

If oxidation of sulfides continues at values pH<4, the generation of AMD occurs, which can be neutralized if the encasing rock has enough carbonates, hydroxides, and silicates to consume the acidity generated [3]. However, if oxidation persists, and pH <3, precipitation of secondary phases such as ferric oxyhydroxides and gypsum takes place, which is accumulate, causing cementation and agglomeration of grains called "hardpan", it decreases the porosity below the surface. The formation of hardpans limits water infiltration and vertical oxygen diffusion [27], for this reason they are considered hydraulic and diffusive barriers that protect the non-weathered material from oxidation [28]. In historical residues from New Zealand [29] they found that hardpand is mainly composed of very fine minerals (μm and nm) of Fe-As-S, in which the oxyhydroxides of As, bukovskyite [$Fe_2(AsO_4)(SO_4)(OH).7H_2O$] and scorodite ($FeAsO_4.2H_2O$) are the most abundant. The formation of cement is facilitated in dry climates that allow the evaporation process that improves the cementation of minerals. Although, hardpans serve as sinks for PTEs, their function is not permanent, since their layers could undergo fracturing, and as consequence the infiltration of oxic surface water can cause oxidation of sulfides [30]. On the other hand, the aging of oxyhydroxides (ferrihydrite to goethite) reduces the adsorption capacity due to the increase in crystallinity [31].

Furthermore, precipitation of secondary minerals such as jarosite has a great synergistic capacity to simultaneously incorporate Pb (II) and As (V) in its structure, during mineral growth and mineral-water interactions; amount of As (V), which replaces SO_4 is greater when Pb (II) is also incorporated, in the same way amount of Pb incorporated in the structure is also greater when As (V) is incorporated, this simultaneity seems to confer less aqueous solubility to jarosite [32]. Despite presence of hardpans, if the oxidation of sulfides continues, acid dissolves mineral species that contain high concentrations of PTEs, and then are available to reach stream water through runoff. PTEs in surface waters are found in their different compartments. However, sediments are considered the main sink and transport medium, since, through adsorption, precipitation, co-precipitation and coadsorption they can remove highly toxic elements; the adsorption and coprecipitation in Fe minerals limit migration of pollutants in aquatic environments [33]. Although [34] have been found by SEM (Scanning Electron Microscope), that johansenite (manganese pyroxene) originates MnO which could be better adsorbent of PTEs than FeO.

In the Yinma River, in Northeast China [35] they found that Pb and Cu had higher adsorption affinity to sediments than Ni and Cd, and they were adsorbed in higher concentrations to Fe and Mn oxyhydroxides, than in matter organic and residual (solid primary minerals). Cu in ferrihydrite is adsorbed by outer sphere interaction weak bonds (ion exchange) and by inner sphere interaction strong bonds (specific adsorption). However, the presence of organic carbon (OC) causes ferrihydrite, although it precipitates at a smaller size (5 nm to 1 nm), forms a layer or cover that inhibits adsorption of Cu, so Cu is contained mainly by coprecipitation, being trapped in cavities, when precipitation takes place [36]. In addition, it have been found that Cu is adsorbed to humic acids, by ionic bonding and complex formation through its carboxylic and phenolic functional groups, sorption capacity is mainly carried out at pH <4, at higher pH, sorption could be complicated [37].

The mobility and fate of As in sediments and groundwater is strongly controlled by the sorption process, and its extent of adsorption is influenced by the presence of OM. Coprecipitation/preadsorption of HA in ferrihydrite inhibits As from binding to Fe oxyhydroxides because OM can compete with As for available binding sites, promoting the mobility of As (v) > As (III) [38]. Likewise, the retention of As (III) and As (V) on goethite surfaces is reduced in the presence of (HA) and (FA) [39]. Furthermore, the sorption of As in sulfurous minerals also influences mobility, since it can co-precipitate in FeS_2 or precipitate in sulphides such as rejalgar (As_4S_4)

[40]. Several studies indicate that Zn and Cd are highly mobile, Zn > Cd; [41] found that Zn presented high correlation with aluminosilicates, and the adsorption results indicate that it is mainly adsorbed on clays through weak external sphere bonds. Aquatic environments commonly contain graphene oxide (a very soluble substance), which is characterized by its surface containing carboxyl groups that can form complexes with metal ions and coadsorb (approx 91%) to hematite and/or goethite at pH 3-5, decreasing its adsorption when it increases; this coadsorption is considered an irreversible process [10].

5. Conclusions

Ancient mining waste contains large amounts of sulfurous minerals with highly and concentrated toxic elements that, exposed to weathering, undergo hydrolysis and oxidation–reduction reactions and form AMD, that dissolves mineral phases and release As, Pb, Cd, Zn, Cu and Fe. AMD cause secondary minerals formation such as clays, gypsum, jarosite, ferrihydrite, hematite, among others; when minerals of ferric oxyhydroxide and gypsum are agglomerated and strongly cemented form hardpans that are important sinks mainly of As. Likewise, jarosite, during its growth, incorporates As (V) and Pb (II) in its structure simultaneously, achieving less aqueous solubility.

Due to AMD, the PTEs by runoff reach rainwater and, due to the prevailing pH, neutral to slightly basic in stream water currents, the highly concentrated metals and As are absorbed into the finest particles of the sediments (clays, oxyhydroxides and OM). Thus, the mobility and fate of As and metals in sediments and groundwater is strongly controlled by the sorption process, and extent of adsorption is influenced by the presence of OM. Metals such as As, Pb, Cd and Cu are adsorbed to oxyhydroxides by specific adsorption, by inner sphere strong bonds, causing reduced mobility. On the other hand, ion exchange, although it also takes place in most metals, is more representative of Zn, by outer sphere weak bonds, that cause high mobility. Cu, in the presence of OC, rather than adsorbed by internal sphere bonds, this is co-precipitated in the cavities of the ferrihydrite. Likewise, in sediments rich in sulfides, As can precipitate as rejalgar (As_4S_4) or co-precipitate in pyrite. The coprecipitated/preadsorbed HA in the ferrihydrite inhibit As binding, promoting the mobility of As (v) > As (III).

The risk that PTEs represent to human health can be inferred by their speciation or fractionation chemical perfil. Although sediments have high concentrations of toxic metals, only those found in fraction 1 and 2, soluble/interchangeable and carbonate, respectively, are those what present greater mobility, toxicity, and bioavailability in aquatic environments. However, the concentrations of As and metals in the distinct fractions could undergo changes mainly due to variations in the pH.

In addition, understanding of the physicochemical processes and mineralogy in tailings deposits could contribute to create more efficient protocol and alternatives to reduce mobility of PTEs in sediments and aquatic environments and consequently reduce effects on human beings.

Acknowledgements

Authors would like to thank Consejo Nacional de Ciancia y Tecnología (CONACYT) for the studentship and support for project number 2121.

Conflict of interest

The authors declare no conflict of interest.

Author details

Marilú Barrera Olivarez[1], Mario Alfonso Murillo Tovar[2*], Josefina Vergara Sánchez[3], María Luisa García Betancourt[4], Francisco Martín Romero[5], América María Ramírez Arteaga[1], Gabriella Eleonora Moeller Chávez[6] and Hugo Albeiro Saldarriaga Noreña[4]

1 Facultad de Ciencias Químicas e Ingeniería, Universidad Autónoma del Estado de Morelos, Cuernavaca, Morelos, México

2 CONACYT-Centro de Investigaciones Químicas-IICBA, Universidad Autónoma del Estado de Morelos, Cuernavaca, Morelos, México

3 Escuela de Estudios Superiores de Xalostoc, Universidad Autónoma del Estado de Morelos, Ayala, Morelos, México

4 Centro de Investigaciones Químicas-IICBA, Universidad Autónoma del Estado de Morelos, Cuernavaca, Morelos, México

5 Instituto de Geología, Universidad Nacional Autónoma de México, Ciudad Universitaria, Circuito de la Investigación Científica, Ciudad de México, México

6 Universidad Politécnica del Estado de Morelos, Jiutepec, Morelos, México

*Address all correspondence to: mario.murillo@uaem.mx

References

[1] Adriano, D.C. (2001). Trace Elements in Terrestrial Environments. *Biogeochemistry, Bioavailability, and Risks of Metals*. Volume I. Second Edition. Springer, 867 pp.

[2] Alloway, Dr, B.J. (2012). Heavy Metals in Soils. (2a ed). John Wiley & Sons, Inc.

[3] Montes-Ávila, I., Espinosa-Serrano, E., Castro-Larragoitia, J., Lázaro, I., Cardona, A. (2018). Chemical Mobility of Inorganic Elements in Stream Sediments of a Semiarid Zone Impacted by Ancient Mine Residues. Applied Geochemistry. Doi.org/10.1016/j.apgeochem.2018.11.002

[4] Lindsay, M, B, J., Moncur, M, C., Bain, J, G., Jambor, J, L., Ptacek, C, J., Blowes, D, W. (2015). Geochemical and mineralogical aspects of sulfide mine tailings. Applied Geochemistry. Doi.org/10.1016/j.apgeochem.2015.01.009

[5] Fergusson, J. E. (1990). The heavy elements. Chemistry, Environmental Impact and Health Effects. Pergamon press.

[6] Romero, F.M., Armienta, M.A., González–Hernández, G. (2007). Solid-phase control on the mobility of potentially toxic elements in and abandoned lead/zinc mine tailings impoundment, Taxco, Mexico. Appl. Geochem. 22, 109-127. Doi.org/10.1016/j.apgeochem.2006.07.017

[7] Sparks, D. L., Scheidegger, A. M., Strawn, D. G. and Scheckel, K. G. (1999). Kinetics and Mechanisms of Metal Sorption at the Mineral-Water Interface. Downloaded by North Carolina State Univ. 28,2012. Publication Date: January 15, 1999. Doi: 10.1021/bk-1998-0715.ch007

[8] McGregor, R.G. & Blowes, D.M. (2002). The physical, chemical and mineralogical properties of three cemented layers within sulfide-bearing mine tailings. Journal of Geochemical Exploration 76. 195-207. Doi.org/10.1016/S0375-6742(02)00255-8

[9] Moncur, M.C., Jambor, J.L., Ptacek, C.J., Blowes, D.W. (2009). Mine drainage from the weathering of sulfide minerals and magnetite. Applied Geochemistry. 24. 2362-2373. Doi.org/10.1016/j.apgeochem.2009.09.013

[10] Sheng, G., Huang, Ch., Chen, G., Sheng, J., Ren, X., Hu, B., Ma, J., Wang, X., Huang, Y., Alsaedi, A. and Hayat, T. (2018). Adsorption and co-adsorption of graphene oxide and Ni (II) on iron oxides: A spectroscopic and microscopic investigation. Environmental Pollution 233. 125-131. Doi.org/10.1016/j.envpol.2017.10.047

[11] Lindsay, M. B.J., Condon, P. D., Jambor, J. L., Lear, K. G., Blowes, D. W., Ptacek. C. J. (2009). Mineralogical, geochemical, and microbial investigation of a sulfide-rich tailings deposit characterized by neutral drainaje. Applied Geochemistry. 24. 2212-2221. Doi.org/10.1016/j.apgeochem.2009.09.012

[12] Petrucci, R.H., Harwood, W.S., Herring, F.G. (2003). Química General. Octava edición. Pearson Educación. P. p. 1288.

[13] Chang, R. (2007). Química. Novena edición. McGrawHill. P.p. 1063

[14] Bao, Z., Bain, J., Holland, S.P., Wilson, D., Mackenzie, P., Ptacek, C.J., Blowes, D. W. (2020). Faro Waste Rock Project: Characterizing geochemical heterogeneity in sulfide-and carbonate-rich waste rock. Applied Geochemistry 121. 104691

[15] Bennett, J.W., Harries, J. R., Ritchie, I. M. (1987). Rehabilitation of waste

rock dumps at the rum Jungle Mine site. Proceedings America Society of Mining and Reclamation. Pp 104-108. Doi: 10.21000/JASMR88010104

[16] St-Arnault, M., Vriens, B., Blaskovich, R., Aranda, C., Klein, B., Mayer, K. U., Beckie, R. D. (2020). Geochemical and mineralogical assessment of reactivity in a full-scale heterogeneous waste-rock pile. Minerals Engineering 145, 106089. Doi.org/10.1016/j.mineng.2019.106089

[17] Lorca, M. E., Mayer, K. U., Pedretti, D., Smith, L., Beckie and R. D. (2016) Spatial and Temporal Fluctuations of Pore-Gas Composition in Sulfidic Mine Waste Rock. Vadose Zone Journal. 15 (10). Doi.org/10.2136/vzj2016.05.0039

[18] Romero, F. M., Armienta, M.A., Gutiérrez, M. E., Villaseñor, G. (2008). Factores geológicos y climáticos que determinan la peligrosidad y el impacto ambiental de jales mineros. Rev. Int. Contam. Ambient. 24(2) 43-54.

[19] Baran, A & Tarnawski. 2015. Assessment of heavy metals mobility and toxicity in contaminated sediments by sequential extraction and battery of bioassays. Ecotoxicology 24:1279-1293. Doi.org/10.1007/s10646-015-1499-4

[20] Tessier, A., Campbell, P.G.C., Bisson, M. (1979). Sequential Extraction Procedure for the Speciation of Particulate Trace Metals. Analytical Chemistry. Vol. 51: 844-851, No.7.

[21] Mitra, S. (2003). Sample Preparation Techniques in Analytical Chemistry. Wiley-Interscence.

[22] Espinosa, E. & Armienta, M. A (2007). Mobility and fractionation of Fe, Pb and Zn in river sediment from a silver and base-metals mining area: Taxco, Mexico. Journal of Environmental Science and Health Part A. 42: 1391-1401. Doi.org/10.1080/10934520701480474

[23] Foster, A. L., Brown, G. E., Tingle, T. N and Parks, G. A. (1998). Quantitative arsenic speciation in mine tailings using X-ray absorption spectroscopy. American Mineralogist, Vol. 83, Pages 553-568. Doi.org/10.2138/am-1998-5-616

[24] Armienta, M.A., Dotor, A., Arcega, F., Talavera, O., Aguayo, A., Ceniceros, N., Cruz, O. (2007). Assessing water–sediment processes for metals in rivers polluted by mining to predict environmental impacts in developing countries. *Water Quality and Sediment Behaviour of the Future: Predictions for the 21st Century.* IAHS Publ. 314.

[25] Loredo, R., Bustamante, J., González, H. N., Moreno, V., Del Río, R., Molina, F., González, B., Archundia, D. (2020). Mobility and accessibility of Zn, Pb, and As in abandoned mine tailings of northwestern Mexico. Environmental Science and Pollution Research. Doi.org/10.1007/s11356-020-09051-1

[26] Khoeurn, K., Sakaguchi, A., Tomiyama, S., Igarashi, T. (2019). Long-term acid generation and heavy metal leaching from the tailings of Shimokawa mine, Hokkaido, Japan: Column study under natural condition. Journal of Geochemical Exploration. 201, 1-12. Doi.org/10.1016/j.ge xplo.2019.03.003

[27] Elghali, A., Benzaazoua, M., Bussière, B., Kennedy, C., Parwani, R., Graham, S. (2019). The role of hardpan formation on the reactivity of sulfidic mine tailings: A case study at Joutel mine (Québec). Science of the Total Environment. 654, 118–128. Doi.org/10.1016/j.scitotenv.2018.11.066

[28] DeSisto, S. L., Jamieson, H. E., Parsons, M. B. (2011). Influence of hardpan layers on arsenic mobility in historical gold mine tailings. Applied Ge ochemistry. 26, 2004-2018. Doi.org/10.1016/j.apgeochem.2011.06.030

[29] Kerr., G & Craw., D. (2020). The role of hardpan formation on the reactivity of sulfidic mine tailings: A case study at Joutel mine (Québec). New Zealand Journal of Geology and Geophysics. Doi.org/10.1016/j.scitotenv.2018.11.066

[30] Even, E., Masuda, H., Shibata, T., Nojima, A., Sakamoto, Y., Murasaki, Y. and Chiba, H. (2017). Geochemical distribution and fate of arsenic in water and sediments of rivers from the Hokusetsu area, Japan. Journal of Hydrology: Regional Studies 9, 34-47. Doi.org/10.1016/j.ejrh.2016.09.008

[31] Ford, R.G., Bertsch, P.M. and Farley, K. J. (1997). Changes in Transition and Heavy Metal Partitioning during Hydrous Iron Oxide Aging. Environ. Sci. Technol. 31, 2028-2033. Doi.org/10.1021/es960824+

[32] Aguilar, J., Villalobos, M., Pi, T., Escobar, I. N, and Romero, F. M. (2018). Synergistic arsenic (V) and lead (II) retention on synthetic jarosite. I. Simultaneous structural incorporation behaviour and mechanism. Environmental Science Processes & Impacts. 20, 354. Doi.org/10.1039/C7EM00426E

[33] Luo, C., Routh, J., Dario, M., Sarkar, S., Wei, L., Luo, D., Liu, Y. (2020) Distribution and mobilization of heavy metals at an acid mine drainage affected region in South China, a post-remediation study. Science of the Total Environment. 724. Doi.org/10.1016/j.scitotenv.2020.138122

[34] Kim, Y. (2020). Geochemical Behavior of Potentially Toxic Elements in Riverbank-Deposited Weathered Tailings and Their Environmental Effects: Weathering of Pyrite and Manganese Pyroxene. Pollutants in Acid Mine Drainage. 10(5), 413. Doi.org/10.3390/min10050413

[35] Guan, J., Wang, J., Wang, Z., Zhuang, J., Dong, D., Luo, M., Lu, N., Yuan., X. (2019). Comparison of Pb(II), Cu(II), Cd(II), and Ni(II) adsorption onto surficial sediment components from aquatic environments in the Phaeozem zone of Northeast China. Journal of Geochemical Exploration 197, 220–227. Doi.org/10.1016/j.gexplo.2018.12.005

[36] Seda, N.N., Koenigsmark, F. and Vadas, T. M. (2016). Sorption and coprecipitation of copper to ferrihydrite and humic acid organomineral complexes and controls on copper availability. Chemosphere 147. 272-278. Doi.org/10.1016/j.chemosphere.2015.12.106

[37] Yang, K., Miao, G., Wu, W., Lin, D., Pan, Bo., Wu, F., Xing, B. (2015). Sorption of Cu^{2+} on humic acids sequentially extracted from a sediment. Chemosphere. 138, 657-663. Doi.org/10.1016/j.chemosphere.2015.07.061

[38] Xue, Q., Ran, Y., Tan, Y., Peacock, C. L., & Du, H. (2019). Arsenite and arsenate binding to ferrihydrite organo-mineral coprecipitate: Implications for arsenic mobility and fate in natural environments. Chemosphere. Doi.org/10.1016/j.chemosphere.2019.02.118

[39] Grafe, M., Eick, M. J., Grossl, P.R., Saunders, A. M. (2002). Adsorption of arsenate and arsenite on ferrihydrite in the presence and absence of dissolved organic carbon. Journal Environ. Qual. 31, 1115-1123. Doi.org/10.2134/jq2002.1115

[40] Jay, J.A., Blute, N.K., Lin, K., Senn, D., Hemond, H. F., Durant, J.L. (2005). Controls on arsenic speciation and solid-phase partitioning in the sediments of a two-basin lake. Environ. Sci. Technol. 39 (23), 9174-9181. Doi.org/10.1021/es051216+

[41] Gunawardana, C., Agodawatta, P., Goonetilleke, A. (2015). Adsorption and mobility of metals in build-up on road surfaces. Chemosphere. 119, 1391-1398. Doi.org/10.1016/j.chemosphere.2014.02.048

Section 3

Methods for Water Quality Improvement

Chapter 8

Physicochemical Insights of the Organic Matter Particles Dispersed in Wastewaters Induced by Bio-Polyelectrolytes

Carlos A. Quintero Gonzalez, Eduardo A. López-Maldonado and Mercedes T. Oropeza-Guzmán

Abstract

The turbidity and color of the water are mainly caused by colloidal particles. These particles remain in suspension for a long time and can even pass through a very fine filter medium, since they do not have a tendency to agglomerate. Due to this, polyelectrolytes such as chitosan have been used in coagulation-flocculation processes because they dissociate into charged species in solution and these contribute to charges or dissociable groups which are covalently bound to its structure. With the zeta potential measurements (ζ) vs. pH and particle size, the ideal dose of bio-polyelectrolyte was determined with which, the isoelectric point (IEP) was reached, generating electroneutrality in the system, removing 92% of the chemical oxygen demand (COD). The results discussed here represent a sustainable alternative to the water reuse and sanitation problem of the fish processing industry. The use of bio-polyelectrolytes offers that the by-products obtained from the coagulation-flocculation process can be reused and recovered for other uses.

Keywords: turbidity, colloidal particles, chitosan, coagulation-floculation, isoelectric point, bio-polyelectrolyte

1. Introduction

The problems of water availability have their origin in the physiographic and climatic distribution, the contamination of surface and underground waters and the accelerated increase in demand for the different uses. For example, the reuse of water in agriculture is a well-known practice in the world; however, reuse in industry, municipal services, secondary uses, and aquifer recharge is only practiced to a limited extent. Given the imminent shortage of the resource in some areas of the country, the reuse of water is considered as an alternative supply; however, to reuse the water it is necessary to meet a certain quality.

Among the main parameters that define water quality are turbidity and color, attributed to the presence of colloidal particles and dissolved organic matter. These particles remain in suspension for a long time and can even pass through a very fine filter medium [1] since they do not have the tendency to agglomerate.

In particular, coagulation and flocculation processes have shown their usefulness in eliminating colloidal particles in water. On one hand, coagulation aims to neutralize the surface charges of suspended particles, and thus facilitate their agglomeration. In practice, this procedure is characterized by the injection and rapid dispersion of both chemical and organic products [2]; on the other hand, flocculation aims to favor, with the help of slow mixing, the contact between destabilized particles. Thus, the particle agglomerate forms a "floc" that can be easily eliminated by the decantation and filtration procedures [3].

Polymers have been used in coagulation-flocculation processes for water purification for at least four decades [4]; compared to alum, some of the advantages derived from the use of polymers in water treatment are:

- use of a lower dose of coagulant.
- less generation of sludge.
- less increase in the ionic charge of the treated water.
- cost reduction of 25–30%.

The polymers used in water treatment are soluble in water and mainly synthetic in nature, although some natural products may be of interest. Polymers are widely characterized by their ionic nature: cationic, anionic, and non-ionic. Strictly ionic polymers are called polyelectrolytes [5].

The most important characteristics of polymeric flocculants are their molecular weight (MW) and, in the case of polyelectrolytes, their charge density. All polymers used as flocculants in water treatment must be soluble in water.

In an aqueous solution, polymers very often assume a random folded configuration. The length of the polymer depends on the interaction between its segments, if there is an appreciable repulsion between segments then the polymer expands a little.

The best-known example is polyelectrolytes, where the segments are charged. In this case, the polymer can expand significantly and there are significant effects on ionic strength. At higher ionic strength, the repulsion between charged segments is "filtered" by ions in solution, so the expansion of the molecule would not be as great [6].

Due to this aspect, the polyelectrolytes are suitable for use in the coagulation-flocculation process and their use as coagulants in water treatment and sludge thickening. Thus, it is proposed to study the influence of bio-polyelectrolytes (BP), in the aggregation of particles of organic matter dispersed in water. The intention is to eliminate the contamination matrix that is generated in fish packing plants that currently lack water treatment systems. As an additional benefit, it is proposed that the residual sludge does not become waste for special handling, but rather has a reuse value taking advantage of its content. In this work, the performance of bio-polyelectrolytes as a sustainable alternative for the sanitation and reuse of treated wastewater is evaluated. The main factors that affect the performance of chitosan in destabilizing and aggregating colloidal particles of organic matter is discussed.

2. Experimental

The experimental protocol was proposed in three stages derived from taking a sample of the residual water generated in a fish packing company. In the first stage, the quality of the wastewater from the fish packing process was evaluated through measurements of chemical oxygen demand (COD), total organic carbon (TOC), total nitrogen (TN), total phosphorus (TP), etc. In addition, the biodegradability of the contaminants contained in water was evaluated by means of a solid model. In the second stage, the coagulation-flocculation windows (ζ vs. pH) were determined and in the third part, the factors of pH, ζ and polyelectrolyte concentration were correlated in order to evaluate the organic matter separation process in the wastewater of the fish packing plant measuring turbidity, COD (Horiba Coda-500-A), COT (Hatch colorimetric measurement), TN (Hatch colorimetric measurement) and TP (NMX-AA-029-SCFI-2010).

2.1 Coagulants-flocculants agents

Polyelectrolytes of both natural and synthetic origin can flocculate colloids. Polyelectrolytes are polymers with a high molecular mass, which normally contain ionizable functional groups. Typical examples of polyelectrolytes are shown in **Table 1**.

In a simpler way, a polyelectrolyte (PE) can be defined as a polymer that dissociates into charged species in solution. Normally, the term polyelectrolyte is used for polymeric systems that consist of macroions, that is, macromolecules that carry charges or dissociable groups covalently bound to their structure, which in turn must be compensated for counter ions of opposite charges to guarantee the electroneutrality of the system [7].

2.2 Measurement of zeta potential (ζ)

In the zeta potential measurement, the Anton Paar Lite sizer 500 equipment was used, in which a small amount of sample (100 μL) is injected into a univette cell containing two electrodes that are used to create an induced electric field. Once the electric field is applied, the particles move towards the anode or the cathode depending on whether the surfaces are positively or negatively charged. The direction of motion indicates positive charge versus negative charge. The speed of the particle's motion is used to calculate the magnitude of the charge.

2.3 Measurement of particle size (DLS)

For the measurement of the particle size, the Anton equipment for lite sizer 500 was used, which uses the dynamic light scattering technique (DLS), which indicates the mean value of the intensity distribution called the "Z average" and the polydispersity index with which the width of the distribution is described.

2.4 Jar-test in fish packing wastewater

The jar technique was carried out in order to determine biopolyelectrolyte dosages among other parameters. In it, the coagulation-flocculation processes were simulated at the laboratory level. Its versatility lies in using a series of jars at the same time with variations in stirring speed (rpm). Chemical and hydraulic factors influence this process. Among these we have:

- pH;
- temperature;
- coagulant concentration;

- sequence of application of substances;
- degree of agitation;
- sedimentation time.

The pH plays a very important role in the study of the coagulation-flocculation phenomena. This is how a part of the charge of the colloidal particles that have absorbed OH- ions remains destabilized in the colloidal suspension.

Organic flocculants	Structure
Manuranic acid	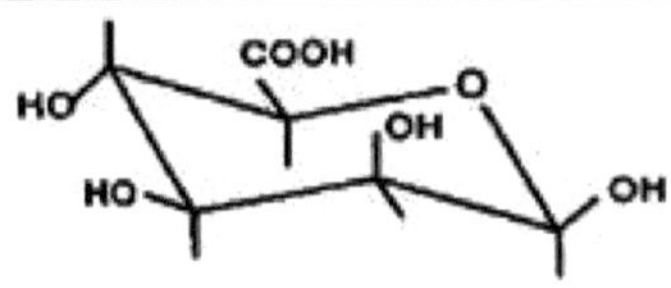
Moringa	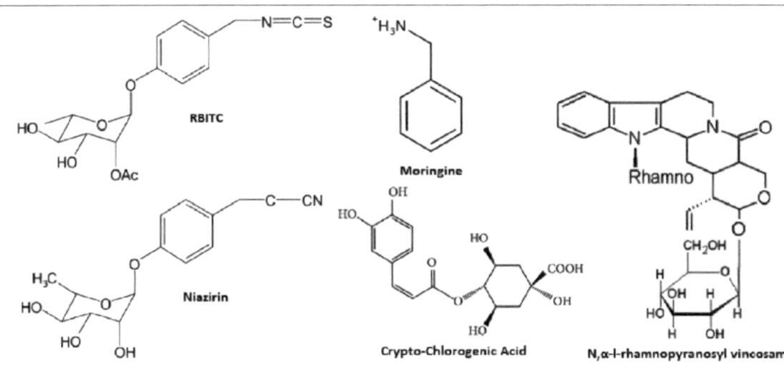
Gluconic acid	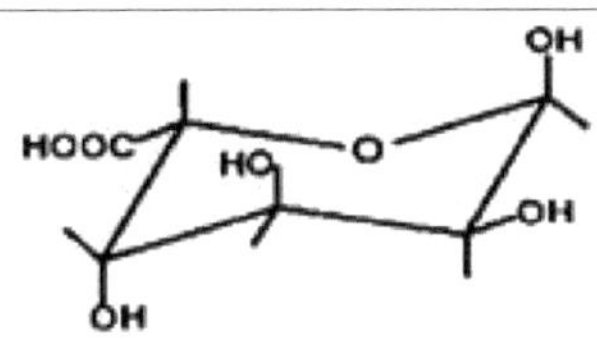
Chitosan	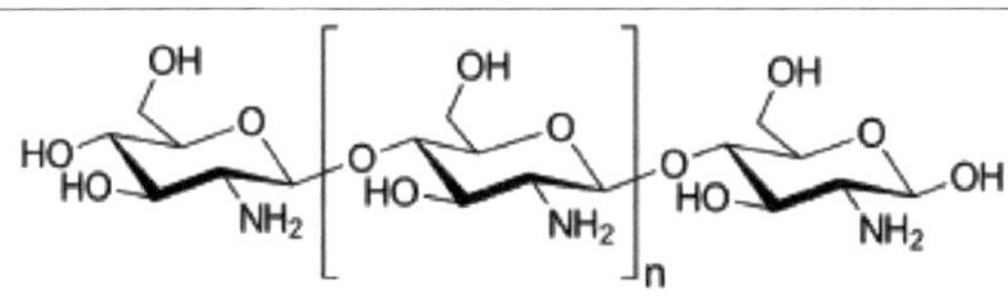
Corn gum	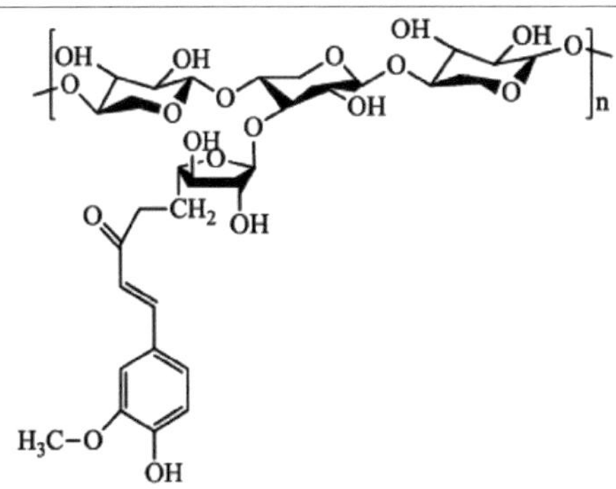

Table 1.
Organic flocculants.

Since the jar test is only a process simulation, it is necessary to maintain operating conditions such as rapid mixing, which aims to create turbulence or movement necessary to create the necessary contact between the bio-polyelectrolyte and the colloidal particles of the water, in order to neutralize their charges, destabilize them and make them agglomerate in a short period of time. The rapid mixing time is directly dependent on the nature of the coagulant. For example, polymers distribute more slowly than metal ions due to their chain length, therefore they will require a longer time or a greater speed gradient.

800 mL of residual water was taken, then additions of 20, 40, 80, and 100 mL of a chitosan solution were made at 1,000 ppm and the time and rpm that were used are 5–10 min, 30–100 rpm [8].

Generally, the slow mixing time does not exceed 15 minutes. Since an excessive time can create heating of the sample causing more efficient flocculation, but at the same time poor sedimentation, the release of gases dissolved in water occurs, forming bubbles that adhere to the flocs and make them float [9]. In this process, it is recommended 3–15 min., 20–40 rpm, finally, a rest of 30 min., which will allow the formed flocs to settle.

3. Results and discussion

The first strategic sampling area was carried out at the entrance of the process with water from the well, taking the NMX-AA-003-SCFI-2019 standard as a sample reference, the total hardness of this water was evaluated in order to identify the presence of salts of calcium and magnesium ions which are responsible for incrustations and can generate failures and loss of efficiency in the processes. In **Table 2**, the results are shown, which indicate that this concentration of salts will not generate any type of interference in industrial processes.

On the other hand, TOC and COD were measured in the wastewater from each process in order to quantify all the carbon present in the organic matter that dissolves or can be suspended in the water and thus, establish the amount of oxygen consumed in the total chemical oxidation of organic constituents present in the water through COD values. Complementary to the characterization of

Parameter	Well water	Waste water sanguaza	Waste water flour process
Temperature	20	4	30
PH	7.017	7.4	6.9
Zeta potential (mV)	1.8	−18.8	16.3
Turbidity (NTU)	12	386	737
Total suspended solids (mg/L)	7	302	517
Total handness (mg/L)	20–50	—	—
COD (mg/L)	—	305	434
TOC (mg/L)	—	209	317
TN (mg/L)	—	306	297
TP (mg/L)	—	284	227
Size particle(nm)	492	1.022	1.196

Table 2.
Physicochemical characterization of wastewater.

the wastewater, the parameters of TN and TP were also evaluated. Because the wastewater exceeds the maximum permissible limits of contaminants in water, a treatment must be carried out which reduces the degree of contamination, since it has a large contribution of essential nutrients for the growth of organisms, such as nitrogen and phosphorus. These can stimulate the growth of macro and photosynthetic microorganisms in harmful quantities, which would trigger a eutrophic environment in water bodies.

On the other hand, the parameter ζ was measured because it is directly related to the ionic properties of the dissolved solids present in the wastewater.

Solid/aqueous interfaces suffer from a pH-dependent surface charge, that is, they tend to have a positive charge at acidic pH and a negative charge at basic pH. The sign and magnitude of the surface charge govern the adsorption of ionic species in solution and the physical properties of dispersions (e.g., their stability against coagulation). That is why, in the coagulation process, the suspended solids are destabilized by varying the pH of the water and this tends to zero when it approaches the isoelectric point. A simple variation of pH could be enough to stabilize or destabilize the solids that are dissolved in the water [10]. Therefore, the pH value can control both the charge density of the bio-polyelectrolyte and the surface charge of the suspended particles (**Figure 1**).

The surface charge of both coagulants and colloids in water depend on pH and their behavior having a great influence and performance in the coagulation-flocculation processes, which is why ζ measurements are required to characterize the colloidal system and understand repulsion and aggregation between colloidal particles [11].

The dosing strategy was determined by measuring the ζ of both the organic matter colloids and chitosan, and also by the optimum pH value to reach the isoelectric point. The purpose of the chitosan dose is to reduce interparticle repulsion by neutralizing negatively charged particles of organic matter.

In general, the electro-neutrality zone for the chitosan-organic matter system is below pH 6 as shown in **Figure 2**. Working at this pH improves the efficiency of the coagulation-flocculation process, since chitosan achieves a higher charge density in an acid medium, therefore its concentration can be decreased.

The best clarification of the wastewater was at pH 5.9, at this pH the positive charges in the chitosan lead to an increase in the hydrodynamic diameter (Dh) due to the repulsion between the intramolecular protonated amino group and this is beneficial in the effect of dosing in turbidity removal efficiency [8]. Furthermore, at this pH, the system reaches the maximum surface contact between chitosan and organic matter in the wastewater [12].

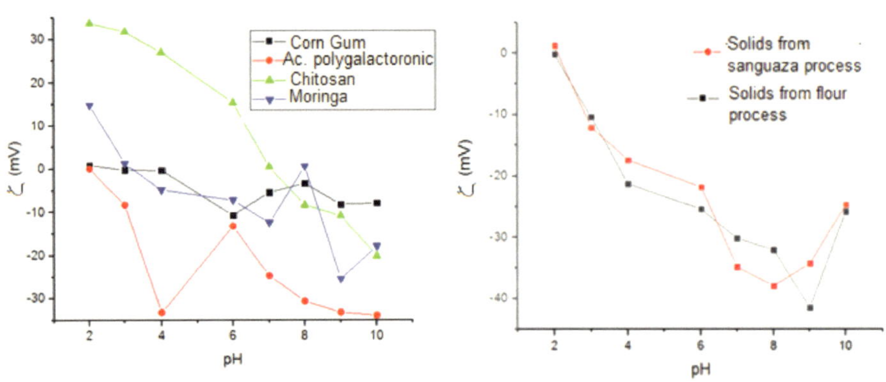

Figure 1.
Ionic behavior of different polyelectrolytes at different pH values.

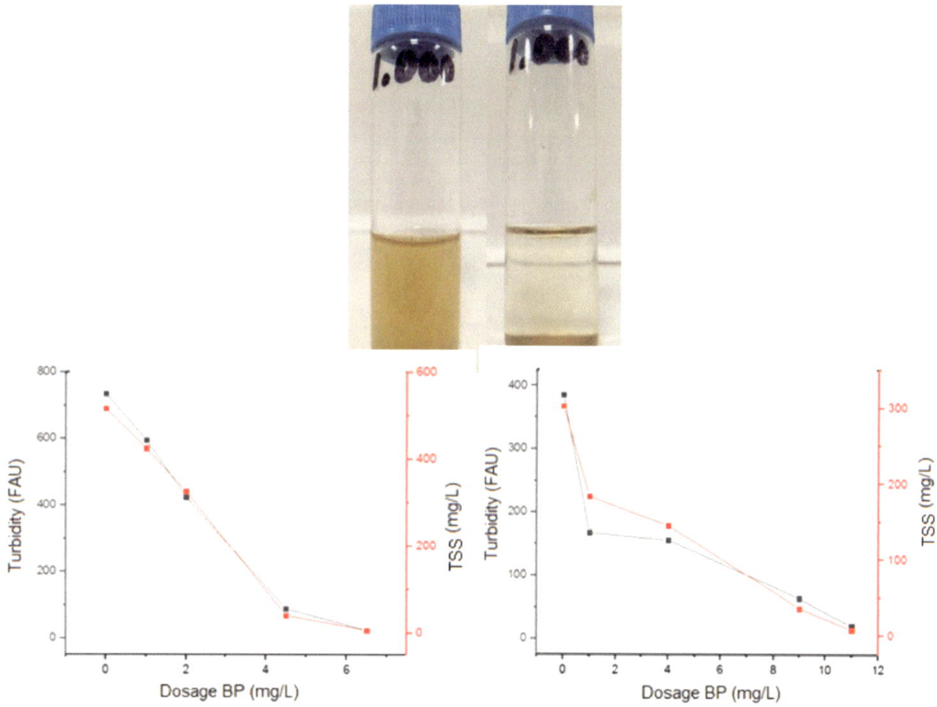

Figure 2.
Effect of chitosan dosage on turbidity removal efficiency. (A) Sanguaza process; (B) flour process. BP: Bio-polyelectrolyte, TSS: Total Suspended Solids.

The optimal dose of chitosan was 10 ppm, thus obtaining a turbidity removal efficiency of around 97% at this dose, the surface charges of the organic matter particles were neutralized by chitosan molecules resulting in a ζ value close to zero.

Particle size influences many properties of particulate materials and is a valuable indicator of their quality and performance. Larger, spherical particles will flow more easily than small ones, but smaller particles dissolve more quickly and result in higher suspension viscosities than large particles. Smaller particle sizes and a high surface charge (ζ) typically improve the stability of the particles in solution [13–15].

For the measurement of the particle size, the dynamic light scattering technique (DLS) was used, as can be seen in **Figure 3**, as the chitosan dose is increased, the size of the particles in solution increases due to the adsorption of organic matter by the active functional groups of chitosan and in turn the decrease in the ζ. This indicates that the system is unstable and that this in turn, as the dose increases, will reach the isoelectric point, generating electroneutrality in the system. Therefore, the suspended solids by gravity and their own weight will settle.

The performance of the coagulation-flocculation process was evaluated with different parameters such as COD, COT and TSS. In addition, different doses of polymer through profiles of ζ and pH were evaluated (**Figure 4**).

Electro-neutrality occurs at $\zeta = -0.3$ at pH = 5.7 and required a dose of 100 ppm of chitosan. At this point, the lowest turbidity was observed in the supernatant (4 FAU) compared to the original turbidity (369 FAU) in the wastewater suspension at room temperature. Furthermore, it can be observed that the floc formation is of higher quality when the isoelectric point is reached. For the COD measurements, the potassium permanganate ($KMnO_4$) procedure was chosen, in which a known amount of KMnO4 is added, which is assessed with a primary calcium oxalate standard. The excess KMnO4 reacts with the excess oxalate and finally, the excess oxalate is titrated with the permanganate and this reaction was carried out at elevated temperature to accelerate the process, obtaining an efficient coagulation-flocculation process, removing a total of 93% of total COD and 100% of TSS in jar tests (**Figure 5**).

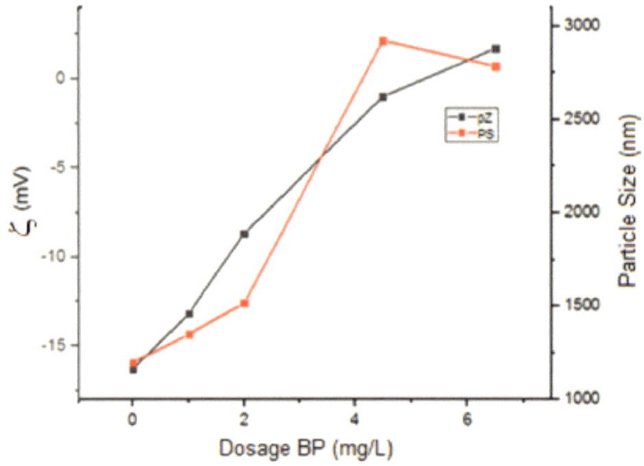

Figure 3.
Effect of chitosan dosage on particle size and Z potential.

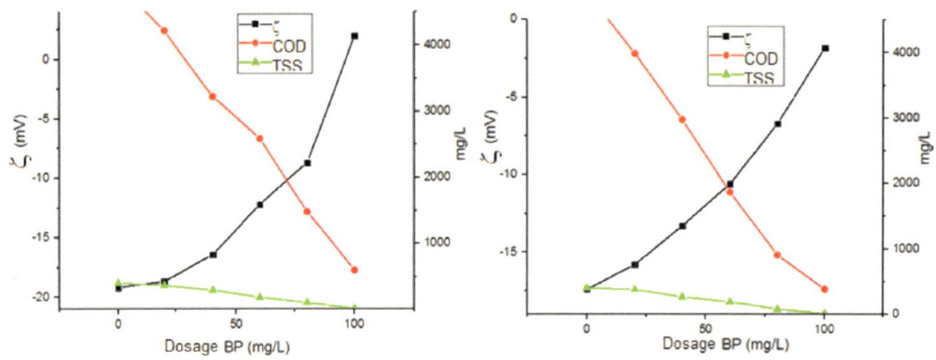

Figure 4.
ζ and evaluation of water quality parameters as a function of chitosan dose in jar tests. (A) Sanguaza process; (B) Flour process. z: zeta potential, BP: Bio-polyelectrolyte, Chemical Oxygen Demand, TSS: Total Suspended Solids.

Figure 5.
Bio-polyelectrolyte contaminant removal efficiencies.

Parameter	Raw water	Filter water	Coagulation-flocculation treatment	% Removal
COD	12,800	4,920	385	92
TOC	2,394	976	198	76
TN	1,854	574	370	35
TP	1,620	487	164	66
TSS	579	397	0	100
Turbidity	595	405	0	100

Table 3.
Evaluation of the quality of the wastewater treated by coagulation-flocculation process.

The main objective of coagulation-flocculation, as it is a primary treatment in wastewater, is to eliminate the suspended solids present in the water as shown in **Table 3**. When carrying out a good dosage of polyelectrolytes, the removal efficiency can be as high as 100%. In the best dose of chitosan, the coagulation-flocculation stage leads to the sedimentation of the suspended pollutants, reaching 92% and 76% removal of COD and TOC, respectively. However, the effluent has a high content of TN and TP represents a risk of eutrophication in the environment, which makes it necessary to couple the coagulation-flocculation process to a secondary treatment.

4. Conclusions

For the treatment of wastewater containing organic matter in suspension, such as a fish packing plant, to be effective, it is necessary to know both the chemical and physical properties of the suspended solids and the water to be treated. The addition of chitosan to wastewater with a high content of organic matter (fish packing plant), triggers a series of processes that begins with the hydrolyzation of the chitosan, followed by the destabilization of the particles by simple specific adsorption of the hydrolysis products and finally sedimentation and separation of organic matter from water. It should be noted that the cationic behavior of chitosan chains allows a better interaction between suspended particles and chitosan itself. Considering the above, the use of biopolyelectrolytes in the coagulation-flocculation process efficiently removes the suspended solids allowing high efficiencies in the removal of COD and TOC, it is worth mentioning that an excess in the dose of bio-polyelectrolyte will re-suspend the settled solids increasing operating costs. This work contributes to the objectives of sustainable development allowing the sanitation and recovery of wastewater using coagulant-flocculating agents that are eco-friendly with the environment.

Acknowledgements

Authors gratefully acknowledge to Consejo Nacional de Ciencia y Tecnología (CONACyT) for the support to Grant No. A1-S-38139. Red Temática de Química Supramolecular proyecto 294810.

Author details

Carlos A. Quintero Gonzalez[1], Eduardo A. López-Maldonado[2]*
and Mercedes T. Oropeza-Guzmán[3]

1 Graduate and Research, Center of the Technical Institute of Tijuana, Tijuana, Mexico

2 Faculty of Chemical Sciences and Engineering, Autonomous University of Baja California, Tijuana, Mexico

3 National Technological Mexico, Tijuana Technological Institute, Chemistry Postgraduate in Engineering Sciences and Postgraduate Science in Chemistry, Mexico

*Address all correspondence to: elopez92@uabc.edu.mx

References

[1] Comisión Nacional del Agua (CONAGUA) (2009), Situación del subsector agua potable, alcantarillado y saneamiento. Edición 2009, México, SEMARNAT, 167p.

[2] Conagua, Semarnat. Estadísticas del Agua en México. Edición 2013. México. 2014.

[3] Jiménez, B.; J.C. Durán, J. M.; Méndez, C. Calidad. En: Jiménez C., M.L. Torregrosa y L. Aboites (Eds.). El Agua en México: cauces y encauces. AMC-Conagua. México. 2014

[4] Acosta-Zamorano, D.; Macías-Carranza, V; Mensoza-Espinosa, L.; Cabello-pasini, A. Agrociencia 2013, 47, 767-779

[5] O.G. Zavala Garcia, M.T. Oropeza-Guzman, W.M. Arguelles Monal, E.A. Lopez-Maldonado, Design and mechanism of action of multifunctional BPÉs with high performance in the separation of hazardous metal ions from polluted water Part I: chitosan-poly(vinylcaprolactam) and its derivatives, Chem. Eng. J. 359 (2019) 840-851.

[6] Edeline, F. Tratamiento fisicoquímico del Agua; Red Hen Press: California, 1997; p 158.

[7] López-Maldonado EA, Oropeza-Guzman MT, Jurado-Baizaval JL, Ochoa-Terán A. Coagulation–flocculation mechanisms in wastewater treatment plants through zeta potential measurements. Journal of Hazardous Materials 2014; 279: 1-10

[8] López-Maldonado, E.A., Oropeza-Guzman, M.T., 2021. Nejayote biopolyelectrolytes multifunctionality (glucurono ferulauted arabinoxylans) in the separation of hazardous metal ions from industrial wastewater. Chem. Eng. J. 423 (28), 130210.

[9] Stuart, M.; Lapworth, D.; Crne, E.; Hart, A. Review of Risk Potencial Emerging Contaminants in UK Groundwater. Sci. Total environ. 2012, 416, 1-21.

[10] Sze, A., Erickson, D., Ren, L. and Li, D., Zeta potential measurement using the Smoluchowski equation and the slope of the current-time relationship in electroosmotic flow, Journal of colloid and interface science, 261, pp. 402-410, 2003.

[11] Feng, X.; Leduc, M.; Pelton, R. Colloids and Surfaces A: Physicochem. Eng. Aspects 2008, 317, 535-542.

[12] Suarez, K; Ponce, S; Lopez, T, Eco-friendly innovation for najayote coagulation-floculation process using chitosan: evaluation through zeta potencial measurements. J. Chemical Engineering. 2014, 536-542.

[13] Sze, A., Erickson, D., Ren, L. and Li, D., Zeta potential measurement using the Smoluchowski equation and the slope of the current-time relationship in electroosmotic flow, Journal of colloid and interface science, 261, pp. 402-410, 2003.

[14] López-Maldonado, E.A., Zavala García, O.G., Cruz Escobedo, K., Oropeza-Guzman, M.T., 2017. Evaluation of the chelating performance of biopolyelectrolyte Green complexes (NIBPEGCs) for wastewater treatment from the metal finishing industry. J. Hazard Mater. 335, 18-27.

[15] E.A. López-Maldonado, M.T. Oropeza-Guzmán, A. Ochoa-Terán, Improving the Efficiency of a Coagulation-Flocculation Wastewater Treatment of the Semiconductor Industry through Zeta Potential Measurements, J. Chem. 2014 (2014) 1-10.

Chapter 9

Performance Evaluation of Waste Materials for the Treatment of Acid Mine Drainage to Remove Heavy Metals and Sulfate

Satish Chandra Bhuyan, Subrat Kumar Bhuyan and Himanshu Bhushan Sahu

Abstract

Acid Mine Drainage (AMD) is the most severe environmental problem facing the mining sector in the current scenario because of low pH and high pollutants concentration. AMD contains a high amount of sulphate viz. pyrite, FeS_2, and to a lesser extent pyrrhotite and heavy metal ions, contaminate both surface water and groundwater. To treat AMD, extensive research projects have been initiated by governments, the mining industry, universities, and research establishments. The environmental impact of AMD can be minimized at these basic levels; prevention should be taken to control the infiltration of groundwater to the pollution site and control the acid-generating process. There are some conventional active methods to treat AMD, such as compost reactor and packed bed iron-oxidation bioreactors; however, these methods have associated with costly material and high maintenance cost, which increases the cost of the entire treatment. In an alternative, the use of low-cost materials such as fly ash, metallurgical slag, zero-valent iron (ZVI), cement kiln dust (CKD), and organic waste such as peat humic agent (PHA), rice husk, and eggshell can be a valuable measure for economic viability to treat the metal-rich wastewater.

Keywords: Acid Mine Drainage, Fly Ash, Metallurgical Slag, Zero Valent Iron, Organic waste

1. Introduction

The extensive mining throughout the globe leads to generate a huge quantity of sulfides caused by weathering process (O_2, moisture, and microbes), which promotes an acidic environment. The static/stream water in contact with mines and mines waste becomes a reaction site where it also acts as a reagent for deferent chemical reactions, i.e., metal/metalloid solubilization. Moreover, in-stream water, it also turns out to be a transport media for reaction products [1]. Such water constituents are called "acid mine drainage" or "acid and metalliferous drainage" (AMD) as shown in **Figure 1**. This water is generally characterized as lower pH value as well as a high heavy metals concentration such as iron, manganese, lead, chromium, mercury, cadmium, and arsenic and sulfate content [2]. AMD has a severe impact on the environment, including neighboring surfaces, groundwater,

Figure 1.
Formation of AMD.

and soil properties. Various reports suggest that heavy metals transmit to the human body cause severe illness and death in AMD-contaminated areas. Therefore, to ensure human health safety and to control the environmental risk, AMD must be treated.

Various alkaline materials such as CaO, NaOH, $CaCO_3$ have been appropriately examined in this chapter. The use of $CaCO_3$ is more than other neutralizing material due to its cost-effectiveness. It produces less amount of sludge but the reaction period is comparably high than other chemical reagents. The major drawback of these alkaline reagents are high in cost and effectiveness in long-term periods is low. To avoid these major drawbacks, there is always a need for a better reagent, which is cheaper, eco-friendly, easy to handle, and the effectiveness in mitigating AMD wastewater. So various industrial by-products are examined by researchers. For example, the by-products generated from the Calcium oxide production process are used to treat sulfate and metals like cobalt, nickel, zing copper cadmium with better efficiency. Some other industrial outcomes like fly ash, steel slag, cement kiln dust, and bayer residue have possible calcium oxide and calcium hydroxide alternatives to treat AMD. The availability of these materials is generally high, which offers cost-effective neutralizing materials for the treatment process.

2. Characteristics of AMD wastewater

AMD is generally characterized by several physio-chemical properties. The chief physical properties are temperature, electrical conductivity, suspended and or dissolved solids whereas the chemical properties are mostly indicated by pH, alkalinity, acidity hardness, the concentration of metal ions, silica, salt, ammonium, and hydrocarbon contents, and radioactivity.

The physical and chemical properties depend on many factors which generally influence the oxidation process of sulfide minerals as well as promotes the migration and dilution of AMD. Hence, every mine has a different material property and should be studied carefully. In the mining sector for metal production, the most crucial factors are pH, the heavy metals concentration, and dissolved anion concentration, i.e., sulfate, chlorides, arsenates, nitrates, etc., and hardness as well as suspended solids. For a better characterization of AMD as well as the properties of waste which are affected by AMD is summarized into five common features listed as follows [3].

i. Acidity, and alkalinity property (pH)

ii. The concentration of different heavy metals

iii. Fe and Al concentration

iv. Sulfate and Arsenate concentration

v. Transparency loss (turbidity and suspended solids)

2.1 Acidity/alkalinity property

The pH is the measure of H^+ concentration; in the case of pure water, H^+ and OH^- are in the same concentration. If the H^+ concentration is higher, it is acidic and if OH^- is higher it becomes alkaline. The pH value of water decreases when it comes in contact with oxidized sulfides. For heavy metals, the lower pH value improves the solubility of the solution and is converted to a toxic solution.

The alkalinity is a measure of the base concentration of a solution and measured by the ability of the solution to absorb protons or the capacity to neutralize a strong acid. It also depends on the CO_2 content and mineralization process. Thus total alkalinity is also known as the sum of OH^-, CO_3^{2-}, HCO^{3-}, NH_3, HS^-, PO_4^{-3}, $H_2BO_3^-$, and organic anions. The measurement of acidity indicates the total acid in the solution; also known as the capacity to neutralize the base.

The solution of a lower pH may contain different abundant acids. For pH higher than 7.0, total acidity rarely exists due to the lower value of sulfuric acid (H_2SO_4). The carbonate/bicarbonates convert into carbonic acid for pH less than 4.2, which then leads to the rapid dissociation into water and CO_2 [4].

2.2 Concentration of different heavy metals

Different heavy metals of high concentrations are the common feature of AMD. Some metals in the metal deposits often incorporate into the AMD at specific geochemical conditions. The toxicity level is a greater problem to aquatic life as well as human health. The control mechanism of heavy metals is quite complex and highly precise to metal and site. In the initial stage where AMD forms, the type of metallic minerals and their solubilities and or dissolution rate control the concentrations of heavy metal. In the second development stage, effluent evolves in contact with regional rocks, atmospheric conditions, and water, changes occur in the complex of metal which favor adsorption and precipitation, so control and mitigate AMD flow [5].

2.3 Iron and aluminum concentration

Iron and aluminum have different geochemical conditions than other heavy metals present in AMD. These two elements are considered seriously due to their higher concentration and effects over other metals. They form a coating along the water stream known as yellow-orange Fe oxy-hydroxides and white-yellow Al oxy-hydroxides as shown in **Figure 2**. For low solubility in nature, these two metals under natural conditions form coloration over the water stream for a long time [6].

Figure 2.
Coloration of AMD water.

2.4 Sulphate and arsenate concentration

Sulphate concentration is the most consistent feature of AMD as its origin comes from sulphate oxidation. Various studies show that the effectiveness of sulphide oxidation depends on sulphate concentration and flow of effluent from the zone of

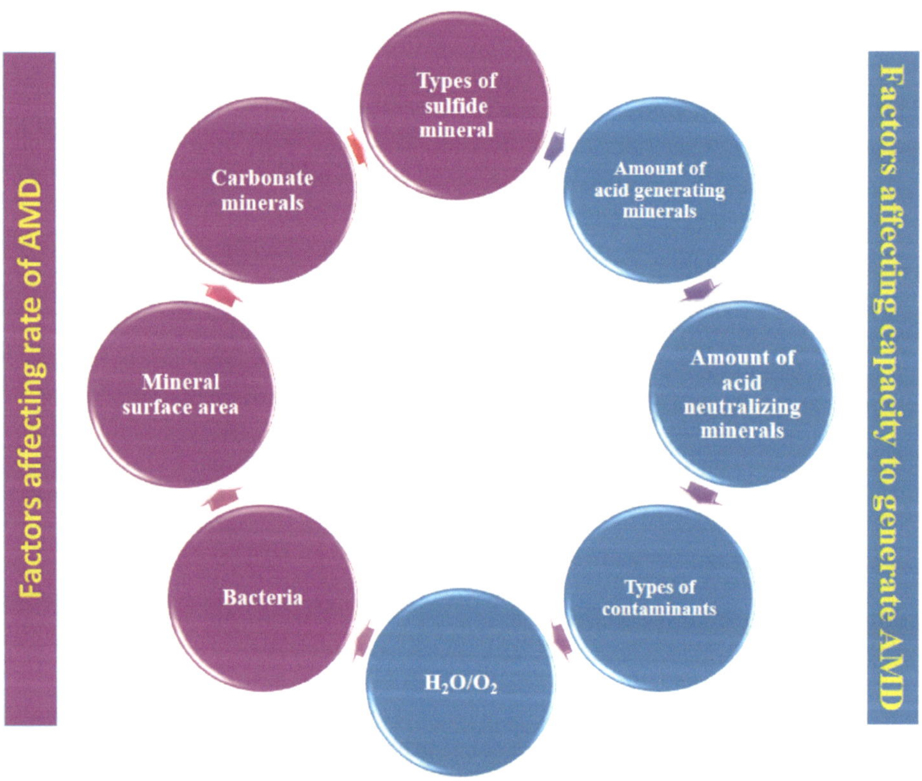

Figure 3.
Various factors for the formation of AMD.

oxidation and any subsequent dilution. Arsenate does not form in such a higher concentration as that of sulphate but this metal has a different area of concern due to its toxic nature [7].

2.5 Turbidity and suspended solids

Turbidity refers to the light absorbance capacity of water preventing its transmission into depth. It is affected by the suspended solids, dissolved solids, and also plankton present in the solution. The measurement of total suspended solids in the laboratory studies indicated various disadvantages related to turbidity; during the storage of these samples, precipitation and flocculation occur. Particles can constitute suspended solid (SS) corresponding to AMD, and macromolecular colloidal particles of aluminum/iron-oxyhydroxides, macroscopic particles, and compounds such as silt and clay. Both the total suspended solids (TSS) and turbidity are of great importance corresponding to the transport phenomena of arsenic and heavy metals in the absorption process, to adverse the water quality and lower down the negative gradient on aquatic life. Also some other factors which is responsible for AMD are shown in **Figure 3** [8].

3. Occurrence of acid mine drainage

AMD normally has a lower value of pH, higher specific conductivity, high concentration of heavy metals such as iron, aluminum, and manganese, and low concentration of heavy metals viz. chromium, nickel, cobalt, arsenic, and so on. The pyrite mineral which is responsible for occurrence of AMD is shown in **Figure 4**. In the current scenario, AMD is left untreated due to inadequate, underdeveloped technologies and or infeasible processes (expensive) in various parts of the globe. The acid generation reaction due to pyrite oxidation, which is widely known as one of the sulphide minerals is given in Eq. (1). The oxidation reaction results in dissolved Fe, sulphate, and hydrogen as reaction products [9, 10].

$$FeS_2 + \frac{7}{2}.O_2 + H_2O \rightarrow Fe^{2+} + 2.SO_4^{2-} + 2.H^+ \tag{1}$$

As the reaction indicated in Eq. (1) moves in the forward direction, the reaction products ferrous iron, sulphate, and hydrogen cataion increase the total dissolved solids (TDS) and hence acidity by lowering pH of solution [11]. If the adjacent surroundings get sufficiently oxidized (depending on oxygen concentration, pH, and microbial activity), much of the Fe^{2+} will be oxidized into Fe^{3+} as expressed in Eq. (2).

Figure 4.
Pyrite mineral.

$$Fe^{2+} + \frac{1}{4}.O_2 + H^+ \rightarrow Fe^{3+} + \frac{1}{2}.H_2O \quad (2)$$

For pH equals 2.3 and 3.5, the ferric iron (Fe^{3+}) precipitates as $Fe(OH)_3$ and [$KFe_3(SO_4)_2(OH)_6$], respectively, a low Fe^{3+} retains in solution which lowers the pH.

$$Fe^{3+} + 3.H_2O \rightarrow Fe(OH)_3\ solid + 3H^+ \quad (3)$$

The leftover Fe^{3+} in Eq. (2) which remains unreacted in Eq. (3) might promote oxidation of additional pyrite as per Eq. (4).

$$FeS_2 + 14Fe^{3+} + 8.H_2O \rightarrow 15.Fe^{2+} + 2.SO_4^{2-} + 16H^+ \quad (4)$$

The aforementioned basic reactions suggest that the acid generation produces ferric iron which gradually precipitates into $Fe(OH)_3$ and may be represented as Eq. (5) which is a combined reaction of Eqs. (1) and (3).

$$FeS_2 + \frac{15}{4}.O_2 + \frac{7}{2}.H_2O \rightarrow Fe(OH)_3 + 2.SO_4^{2-} + 4H^+ \quad (5)$$

In another way, Eq. (6) represents the overall reaction for stable Fe^{3+} used to oxidize additional pyrite.

$$FeS_2 + \frac{15}{8}.O_2 + \frac{13}{2}.Fe^{3+} + \frac{17}{4}.H_2O \rightarrow \frac{15}{2}.Fe^{2+} + 2.SO_4^{2-} + \frac{17}{2}.H^+ \quad (6)$$

In all of the above equations except Eqs. (2) and (3), the oxidant and oxidized mineral are presumed as oxygen and pyrite, respectively. However, pyrrhotite and chalcocite minerals contain altered proportions of metal sulfide and also metals excluding iron [12].

When the water is adequately acidic, acidophilic microbes that flourish at low pH can build up themselves. The microorganism "Thiobacillus Ferroxidans" is assumed to take a huge part in accelerating the synthetic response occurring in mine water circumstances, i.e., these microbes catalyze the oxidation of Fe^{2+}. Another microorganism "Ferroplasma Acidarmanus" has recently been found to play an important role in acid generation in the source water.

Although the formation of H^+ as a result of certain metals precipitations expressed in Eqs. (7) and (8) are not the major acidity sources, these also are considered as treatment alternatives [13].

$$Fe^{+3}/Al^{+3} + 3.H_2O \rightarrow Fe(OH)_3/Al(OH)_3 + 3.H^+ \quad (7)$$

$$Fe^{+2}/Mn + +0.25O_2\ (aq.) + 2.5H_2O \rightarrow Fe(OH)_3/Mn(OH)_3 + 2H^+ \quad (8)$$

Different metals are normally found in AMD because they are available in rocks, like pyrite. There are different metal sulphides viz. ZnS, PbS, NiS, CdS, CuS, etc. which may deliver metal particles into solution but may not produce acidity. The key factors determining the acid generation rate are as follows.

i. Water pH and environment temperature

ii. Oxygen concentration aqueous solution

iii. Saturation degree (in water)

iv. Chemical activity of ferric iron

v. Exposed surface area

vi. Chemical energy (activation) to initiate acid generation

vii. Presence of bacteria that promote oxidation

In the special case where microbial acceleration is significant, some other factors such as activation energy (biological), population density (microbes), and growth rate determine the activity of bacterial. The growth rate depends on pH, temperature, and the presence of various nutrients like nitrate, potassium, ammonia, phosphorous and CO_2 content.

4. Environmental impact of acid mine drainage

AMD shows unique characteristics because the formation and generation of acidic water continue even after the mining area is ceased. So this acidic nature of water is not suitable for the human, animal as well as aquatic life. This problem of the acidic nature of water is not restricted to a limited area near the source of generation but extended to a larger area if this water gets discharged to the main water stream [14]. The AMD has generally more impact on the groundwater than that of the quality of surface water [15]. If the mines which are producing acidic water are present in the permeable formation, this acidic water (low pH)

Heavy metal	Effect on human health	Effect on plant physiological	Permissible level (mg/L)
Cu	Anemia, liver and kidney damage	Inhibits photosynthesis and reproductive process	0.10
Cd	Renal dysfunction, lung infection, and cancer	Decreases seed germination and lipid content	0.01
Zn	Damage to the nervous system	Reduces Ni toxicity, promotes plant development	5.0
Ni	Allergic contact dermatitis, chronic bronchitis, lung, and nasal cancer		0.05
As	Bronchitis, Skin and bladder disease, kidney damage, bone marrow depression	Growth inhibition, Loss of yield and fruit production, Food chain harming	0.05
Mn	Affect the central nervous system	Decreases seed germination, protein, and enzyme	0.1
Hg	Impaired neurodevelopment, decrease in memory	Decreases photosynthetic activity, water uptake, and antioxidant	0.001
Pb	Mental retardation in children, developmental delay	Reduces chlorophyll production and development of plant	0.10
Cr	Affect the nervous system, fatigue, and irritability	Decreases enzyme activity, development of plant; Causes membrane damage, chlorosis, and root damage	0.05

Table 1.
Influence of heavy metals on animals and plants.

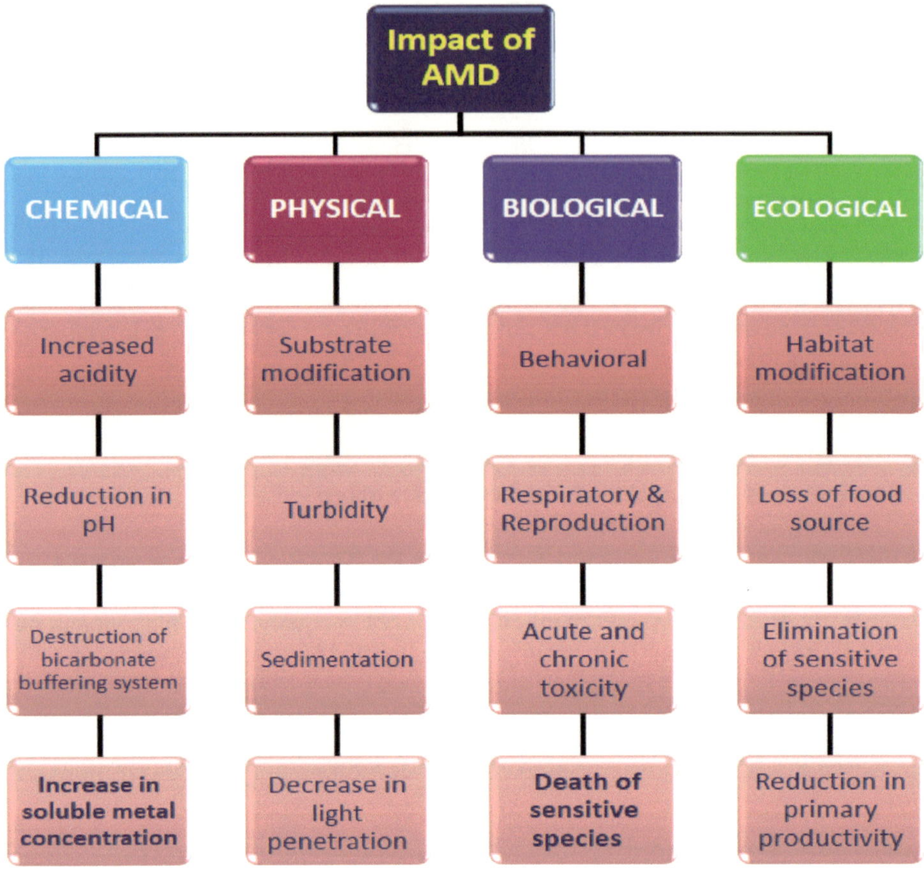

Figure 5.
Several impacts of AMD.

penetrates the aquifer and spreads into and over a larger area with the movement of groundwater which is used by human beings in different ways like wells and bore wells. This acidic and Fe-contaminated water is not only the key reason for corrosion in mine plant equipment by forming scales on the delivery pipe but also pollute the mining atmosphere and surrounding ecosystem [16].

AMD has a serious impact on human heal as well as the ecosystem due to the presence of heavy metals which are not degradable and causes various diseases and disorders in living organisms as well as the plant physiology which is given in the **Table 1**. Various impact of AMD has been shown in **Figure 5**. The pH of this water is very low so the heavy metals present in this water are insoluble and its high concentration causes a toxicological effect on aquatic life. This high presence of metals in water can kill the organisms directly and in the long term effect, lowers the growth and reproduction rates.

5. Treatment technology for AMD

Treatment of AMD is broadly classified into active and passive treatments which is clearly shown in **Figure 6** [17]. The active method is more complicated and required more unit measures and the operational cost is high than other treatment

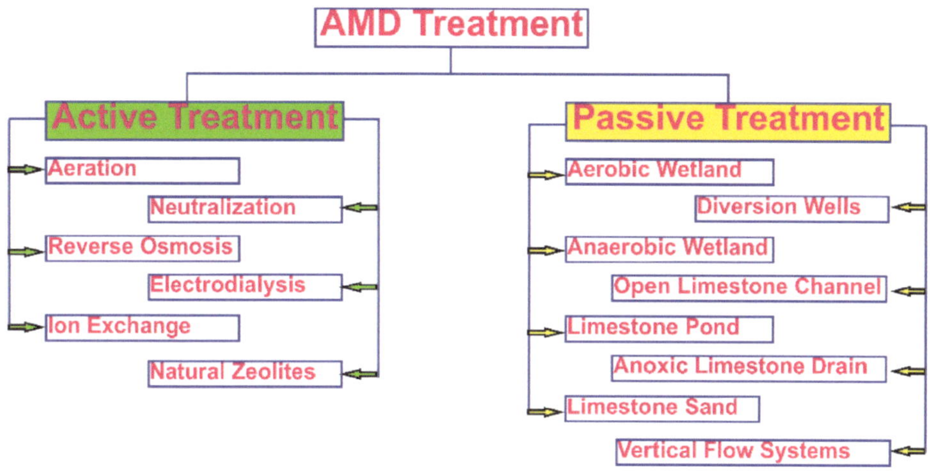

Figure 6.
AMD treatment methods.

methods like passive treatment. The various strategy applied by the various nations for the AMD treatment comprising precipitation, neutralization, ion exchange, electrochemical remediation, adsorption oxidation, etc. All these methods are used for wastewater treatment among which the adsorption technique is the most suitable method for being comparatively economically viable and eco-friendly [18].

5.1 Active technologies for the treatment of AMD

The appropriate strategy used to mitigate wastewater (acidic) that includes the expansion of a chemically neutralizing agent is known as active treatment. The addition of various basic materials will improve the pH and accelerate the oxidation process (chemical) of (Fe^{2+}) and precipitation of different metals into hydroxide and carbonates. The different neutralizing agents viz. lime, slaked lime,

Figure 7.
Common neutralizing reagents.

sodium-carbonate/hydroxide, calcium/magnesium-carbonate/hydroxide can be utilized which is shown in **Figure** 7; differ in cost and efficiency. Although active chemical treatment has better efficiency for mitigation of AMD water, it has a disadvantage of high operational cost and produces more amount of sludge as an effluent. In this treatment process a variety of refinements to improve its efficiency and minimizes the sludge-related problems. The different flocculating reagent is added to increase precipitation. The iron-rich sludge created by the expansion of basic material is highly voluminous and rich in water. This sludge is used as a recycling process by dewatering the sludge in a lime-holding tank [19].

The cost for AMD treatment depends on the location and different scenarios like requirements of manpower, mechanical equipment, and dispensing facilities, and also cost. Soda ash, caustic soda, and ammonia have the lowest cost due to the less expensive equipment needed for the operation. Calcium oxide has the highest cost of installation because of the construction of a lime treatment plant and an aerator. The calcium hydroxide cost is low but the operation cost is high which is used for a long period of treatment for high loading and high acidic condition. Different active technology method which are generally followed for the treatment of AMD is given in **Table** 2.

Sl. No.	Methodology	Membrane used	Feed solutions	Experimental condition	Percentage removal
1.	Reverse osmosis	Cellulose acetate membrane	Fe, Cu, Zn, Ca, Mg, Mn, Ni, and A1	Product rate 26.2 g/hr. for an effective surface area of 13.4 cm^2	95–99% metal separation efficiency [20]
2.	Reverse osmosis	Polyamide ultra-low-pressure reverse osmosis		Feed rate at 1200 L/hr. 0.9–1.0 MPa	Removal percentage of Ni^{2+}, Cu^{2+}, Zn^{2+} and Pb^{2+} was 97.41%, 97.73%, 97.89% and 98.06% respectively [21]
3.	Filtration	Nano-filtration membrane		Feed rate at 1200 L/hr. 0.9–1.0 MPa	Removal percentage of Ni^{2+}, Cu^{2+}, Zn^{2+} and Pb^{2+} was 92.45%, 93.24%, 94.37% and 95.19% respectively [21]
4.	Ion exchange	—		Gel type strong acidic cation exchange resin of the sulphonated polystyrene, porous medium base anion exchange resin with an acrylic matrix	100% Metal removal and 98% of water recovery [22]
5.	Electro-dialysis	HDX 200 anion-exchange and cation exchange membrane		—	The metal removal efficiency was 97% [23]
6.	Natural zeolites	—		6 hrs. of reaction time with a dose of 37 g/L.	Removal efficiency of Fe^{3+}, Mn^{2+}, Zn^{2+}, and Cu^{-} was 80%, 95%, 90%, and 99% respectively [24]

Table 2.
Different active treatment technology.

5.2 Passive technologies for the treatment of AMD

The passive treatments of AMD rely upon biological, physical, and geochemical cycles to improve the nature of water. Primary passive methods can be compressively separated into biological and geochemical systems/reactors that use inorganic substances such as carbonates. The biological systems contain anaerobic and vertical flow wetlands, bioreactors. The geochemical systems include limestone drains, open limestone channels, limestone/steel slag leach beds, limestone sand. The selection of an effective passive treatment method relies on the water chemistry, flow rate, local topography, and characteristics of the site [25].

5.3 Treatment using various waste materials

Active treatment methods are adopted in a wide range but they cause high establishment and absorbent expense. Also, some treatment methods associated with the active process like reverse osmosis, ion exchange requires pre-treatment of influent which is mentioned in the **Table 2**. Waste materials are generally economical than any other treatment method also reduces the environmental load. Various waste material which are generally used for the Treatment process of AMD are given in **Table 3**. They have the capacity and effectiveness to improve the pH of the AMD water and also to remove various pollutants from the wastewater. These waste materials provide a larger surface area, increasing the pH and adsorption rate to remove various pollutants at different concentrations [34, 35].

5.3.1 Fly ash

Fly ash is an unconventional, eco-friendly, low-cost material used as an alternative absorbent for activated carbon. Various research has also suggested that it has also an alternate material to dolomite and limestone used for the pre-treatment process [36]. This waste is used for the treatment of AMD which is successfully removed various heavy metals viz. manganese calcium, iron, aluminum, cadmium, cobalt, zinc, nickel respectively; increases the pH of the solution [26]. The efficiency of fly ash depends on the fly ash characteristics viz. the concentration of CaO and MgO. Initially, the concertation of Ca and Mg increases due to the leaching effect of Ca^{2+} and Mg^{2+} from fly ash surface. But later when gypsum is formed (made of oxygen, sulfur, calcium, and water), concentration Ca and Mg decreases. As the water starts evaporation, it does not protect the sulfur and oxygen-sulfur bonding forms a sulfate (SO_4^{2-}). The sulfate then bonds with calcium (Ca) and water (H_2O) to create gypsum and the Ca concentration decreases. Formation of gypsum occurs at pH >5.5 and absorbs sulphate with high concentration by Fe $(OH)_2$ at pH > 6. Fly ash not only treats heavy metals but also helps in the removal of radioactive material from mine water. It acts as a sink property for the degradation of heavy metals like uranium and thorium. The free CaO present in fly ash attributes to the sulphate removal rate, precipitates gypsum.

5.3.2 Biomass ashes

The synergistic solution is turning out to be exceptionally attractive for sustainability and circular economy where the waste from one industry becomes an asset for another industry, Biomass ash, which is a result of consuming biomass in a power station, can be considered as an effective material for the treatment of acidic water. This ash is a complex alkaline mixture with poly-component, heterogeneous and different variety of composition. Biomass burning is a significant part of

Sl. No.	Used material	Optimum mixing ratio/dose	Initial pH	Final pH	Reaction time	The percentage removal of metals
1.	Fly ash	1:3	2.78	9.1	1440 mins	Greater than 90% for toxic metals, 78% for sulfate [26]
2.	Coal fly ash	—	4	7.0	12 hrs.	60.4% sulfate, 53.4% chemical oxygen demand and removal of Cd^{2+}, Cu^{2+} and Zn^{2+} were 42.9%, 74.8% and 26.7% [27]
3.	FA followed by seeding with gypsum crystals and the addition of amorphous $Al(OH)_3$	1:2	6.6 ± 0.21	12.25	—	Removal of 79.57% sulfate [28]
4.	Alkaline ash leachates	—	3.3–5.0	8.0	7 days	Removal of 99.97% of Cu, 99.78% of Zn, 90.2% of Cd, 99.94% of Pb, 62.71% of Ni, and 99.41% of Co [25]
5.	Modified fly ash	120 g/L	1.6	2.8–6.6	180 min	89%, 92%, 94%, 96%, 60%, and 99% for Ni, Zn, Pb, Fe, Mn, and Al respectively [29]
6.	BOF/SAF slag	30 g/L	2.03	6.32	24 hrs.	Greater than 90% removal of heavy metals and anions [30]
7.	Stainless steel slag	100 gm/L	2.5	5.9	240 mins	Removal of 63.6% iron,39.8% sulfate [31]
8.	Iron slag	30 g/L	2.03	6.68	24 hrs.	Greater than 90% removal of heavy metals and anions [30]
9.	Cement kiln dust slurry	25% of CKD slurry	(2.4 ±0.1)	9.5	1 minute	98% of zinc and 97% of iron [32]
10.	Rice husk	1:10	2.3	4.0	24 hrs.	99% Fe3+, 98% Fe^{2+},98% Zn^{2+}, 95% Cu^{2+} [33]
11.	Peat humic agent	1:500	2.7	3.1	1 hr	Removal of 36% Fe, 26% Al, 20% Zn, 35% Cu, 43% Cd, 98% Pb,40% Ni, 21% Co [34]

Table 3.
Removal of metals using different waste material.

worldwide eco-friendly power which is developing very rapidly overall. Research suggested that the amount of biomass ashes created around the world is 480 Mtpa which can be compared to the coal ash, i.e., 780 Mtpa. Both coal and biomass ashes are generally alkaline and their pH ranges between 9 and 12. They are different in composition such as coal consist of oxides of silicon, aluminum, and iron and content less mount of calcium oxide also the presence of some heavy metals.

5.3.3 BOF and SAF slags

Basic oxygen furnace (BOF) and submerged arc furnace (SAF) slags have complex physical and chemical characteristics [30]. These materials are composed of oxides of calcium, silicon, phosphorus, sulfur, and manganese produced in steel refining processes. Depending upon the sufficiency of the cleaning cycle, inadequately Fe and entrapped metal droplets during the tapping also incorporate into

BOF/SAF slags. Because of the popularity of steel, which is linked to the increase in population results during the production cycle of steel produces more amount of BOF and SAF slags from the production process. Various studies suggested that every year steel industry produced is about 100–200 kg of slag as by-products. After the production, the slags are partially reprocessed but a major part of the slag is used as a landfill material and holds pond and lagoon due to the less demand in the market. This slag raises the pH of AMD and reduces the chemical elements (pollutants) to the desired amount. However, it partially removes the sulphate and also various metals such as Mn, Ni, Co, Zn, Mg. This slag is an ideal candidate for the treatment of AMD and also minimizes the environmental impact with the disposal process of these slags [37].

5.3.4 Eggshell waste

Now a day's egg production rates are higher in various countries so that the waste shell produces from eggs are increasing rapidly [38]. According to the study, the global egg production rate is approaching 86.8 Mtpa globally per year soon. This material is used as alternative treatment material to treat AMD. The primary constituent of ES is $CaCO_3$ is an alkaline material that reacts with acidic water to neutralize it by the process of adsorption and precipitation of metals and also used for the complex, binding, and ion exchange material for various metals ions in the wastewater (Equation 9). This eggshell waste is a very cheap and biodegradable material which are collected, characterized, prepared, and evaluated for the degradation of various anions like aluminum, iron, manganese, and anions like sulphate present in AMD. Due to similar properties like limestone, it can be a good neutralizing agent [39].

$$CaCO_3 + H_2SO_4 \rightarrow CaSO_4 + CO_2 + H_2O \tag{9}$$

6. Discussion

In the above, various discussions have been made for remediation of AMD wastewater, emphasizing waste materials used for the treatment process. Prevention measures play an essential role in controlling AMD formation, but the plan should be made for proper treatment if it can exceed a dangerous level. It is often not wholly possible, requiring corrective techniques to reduce or remove contamination from water. Active and passive treatment methods are mainly utilized to mitigate AMD and improve the quality of the water. Still, their maintenance cost is high, slow treatment process, and requires a large area for operation. But the waste materials play an essential role in mitigating AMD. Different factors like surface area, pH increasing ability, leachability of the material, retention time, cost factor, and environmental impact must be considered when choosing a waste material. The finding results of these waste materials are described below:

- Fly ash and cement kiln dust rich in lime content have better efficiency in removing heavy metals than any other waste material with different pH values.

- Iron and steel slag have ion exchange and sorption properties to degrade metals from the liquid solution.

- BOF and SOF slag are given the same results to increase the pH value above 8 of the aqueous solution; as a result, precipitation of metals occurs.

- The surface area and pH of modified fly ash have more than the fly ash. It requires more reaction time and dose for the absorption process.

- Rice husk was found to be a better reagent and can grow the D. nigrificans, also known as sulfate-reducing bacteria.

- The peat humic agent is also used as an alternative to treat wastewater and can modify kaolinite clay to increase the sorption property of the clay, which can absorb the heavy metals within a pH range of 5 to 8.5.

- Eggshell waste can remove the aluminum and iron content at low temperatures but required a high temperature to remove manganese.

7. Conclusions

This chapter concluded that the demand of various waste materials to be tasted due to their characteristics to determine the suitable condition and amount of dose required to remove metals by absorption process from the AMD. This process also generates some new waste streams and some waste materials that cannot completely remove heavy metals from mine wastewater. So further research and innovation are required to address this issue associated with AMD.

Acknowledgements

We would like to thank National institute of Technology, Rourkela for the support.

Conflict of interest

The authors declare no conflict of interest. The authors declare that they have no known competing financial interests or personal relationships that could have appeared to influence the work reported in this paper.

Author details

Satish Chandra Bhuyan[1]*, Subrat Kumar Bhuyan[2] and Himanshu Bhushan Sahu[1]

1 Department of Mining Engineering, National Institute of Technology, Rourkela, Odisha, India

2 Department of Mechanical Engineering, G.H. Raisoni Institute of Engineering and Technology, Pune, Maharashtra, India

*Address all correspondence to: satishbhuyan2015@gmail.com

References

[1] S Saha AS. Review on treatment of acid mine drainage with waste materials: A Novel Approach. Glob NEST J. 2018 Sep;20(3):512–528.

[2] Jones SN, Cetin B. Evaluation of waste materials for acid mine drainage remediation. Fuel. 2017 Jan 15;188: 294–309.

[3] Favas PJC, Sarkar SK, Rakshit D, Venkatachalam P, Prasad MNV. Acid mine drainages from abandoned mines. In: Environmental Materials and Waste. Elsevier; 2016. p. 413–462.

[4] Koski RA, Munk LA, Foster AL, Shanks WC, Stillings LL. Sulfide oxidation and distribution of metals near abandoned copper mines in coastal environments, Prince William sound, Alaska, USA. Appl Geochemistry. 2008 Feb 1;23(2):227–254.

[5] Soltani N, Moore F, Keshavarzi B, Quality RS-W, and E, 2014 undefined. Geochemistry of trace metals and rare earth elements in stream water, stream sediments and acid mine drainage from Darrehzar Copper Mine, Kerman, Iran. Springer [Internet]. [cited 2021 Jun 21]; Available from: https://link.springer.com/content/pdf/10.1007/s12403-014-0114-x.pdf

[6] No Title. FicMin WH, Plurnlee GS, Smith KS Geol geochemical Control Compos water draining from Divers Miner Depos Cart LMH, Toth, MI, Day, W USGS Res Miner Resour U S Geol Surv Circ 1994;1.

[7] Schrniermund RL, Drozd MA. Acid Mine Drainage and Other. 1997;599. Available from: https://www.worldscientific.com/doi/pdf/10.1142/9781860943768_0013

[8] Hydrometallurgy DJ-, 2006 undefined. Biohydrometallurgy and the environment: intimate and important interplay. Elsevier [Internet]. [cited 2021 Jun 21]; Available from: https://www.sciencedirect.com/science/article/pii/S0304386X06000855

[9] Pozo-Antonio S, Puente-Luna I, Lagüela-López S, Veiga-Ríos M. Techniques to correct and prevent acid mine drainage: A review. DYNA. 2014 Aug;81(186):73.

[10] Akcil A, Koldas S. Acid mine drainage (AMD): Causes, treatment and case studies. Vol. 14, Journal of Cleaner Production. 2006. p. 1139–1145.

[11] Simate GS, Ndlovu S. Acid mine drainage: Challenges and opportunities. J Environ Chem Eng. 2014 Sep;2(3): 1785–1803.

[12] Ben Ali HE, Neculita CM, Molson JW, Maqsoud A, Zagury GJ. Salinity and low temperature effects on the performance of column biochemical reactors for the treatment of acidic and neutral mine drainage. Chemosphere. 2020 Mar;243:125303.

[13] Yan S, Cheng KY, Morris C, Douglas G, Ginige MP, Zheng G, et al. Sequential hydrotalcite precipitation and biological sulfate reduction for acid mine drainage treatment. Chemosphere. 2020 Aug;252:126570.

[14] Vandeginste V, Cowan C, Gomes RL, Hassan T, Titman J. Natural fluorapatite dissolution kinetics and Mn2+ and Cr3+ metal removal from sulfate fluids at 35°C. J Hazard Mater. 2020 May 5;389.

[15] Moodley I, Sheridan CM, Kappelmeyer U, Akcil A. Environmentally sustainable acid mine drainage remediation: Research developments with a focus on waste/by-products. Miner Eng. 2018 Sep;126:207–220.

[16] Skousen JG, Ziemkiewicz PF, McDonald LM. Acid mine drainage formation, control and treatment: Approaches and strategies. Extr Ind Soc. 2019 Jan;6(1):241–249.

[17] Genç H, Tjell JC, McConchie D, Schuiling O. Adsorption of arsenate from water using neutralized red mud. J Colloid Interface Sci. 2003 Aug 15;264 (2):327–334.

[18] Apiratikul R, Pavasant P. Sorption of Cu2+, Cd2+, and Pb2+ using modified zeolite from coal fly ash. Chem Eng J. 2008 Oct;144(2):245–258.

[19] Qiu W, Zheng Y. Removal of lead, copper, nickel, cobalt, and zinc from water by a cancrinite-type zeolite synthesized from fly ash. Chem Eng J. 2009 Jan;145(3):483–488.

[20] Sastri VS. Reverse osmosis separation of metal ions in acid mine-water. Sep Sci Technol. 1979 Oct 1;14 (8):711–719.

[21] Zhong CM, Xu ZL, Fang XH, Cheng L. Treatment of acid mine drainage (AMD) by ultra-low-pressure reverse osmosis and nanofiltration. Environ Eng Sci. 2007 Nov 1;24(9): 1297–1306.

[22] Feng D, Aldrich C, Engineering HT-M, 2000 undefined. Treatment of acid mine water by use of heavy metal precipitation and ion exchange. Elsevier [Internet]. [cited 2021 Jun 24]; Available from: https://www.sciencedirect.com/science/article/pii/S0892687500000455

[23] Buzzi D, Viegas L, ... MR-M, 2013 undefined. Water recovery from acid mine drainage by electrodialysis. Elsevier [Internet]. [cited 2021 Jun 24]; Available from: https://www.sciencedirect.com/science/article/pii/S0892687512002750

[24] Motsi T, Rowson N, Mineral MS-IJ of, 2009 undefined. Adsorption of heavy metals from acid mine drainage by natural zeolite. Elsevier [Internet]. [cited 2021 Jun 24]; Available from: https://www.sciencedirect.com/science/article/pii/S0301751609000349

[25] Bäckström M, Sartz L. Mixing of acid rock drainage with alkaline ash leachates—Fate and immobilisation of trace elements. Water, Air, Soil Pollut. 2011 Nov;222(1–4):377–389.

[26] Gitari WM, Petrik LF, Etchebers O, Key DL, Okujeni C. Utilization of fly ash for treatment of coal mines wastewater: Solubility controls on major inorganic contaminants. Fuel [Internet]. 2008 Sep;87(12):2450–2462. Available from: https://linkinghub.elsevier.com/retrieve/pii/S0016236108001075

[27] Jianping X, Haitao W. Study on the Treatment of Acid Mine Wastewater by Coal Fly Ash [Internet]. atlantis-press.com. 2014 [cited 2021 Jun 24]. Available from: https://www.atlantis-press.com/proceedings/icmaee-14/11479

[28] Madzivire G, Maleka P, ... VV-J of environmental, 2014 undefined. Fate of the naturally occurring radioactive materials during treatment of acid mine drainage with coal fly ash and aluminium hydroxide. Elsevier [Internet]. [cited 2021 Jun 24]; Available from: https://www.sciencedirect.com/science/article/pii/S0301479713007342

[29] Sahoo PK, Tripathy S, Panigrahi MK, Equeenuddin SM. Evaluation of the use of an alkali modified fly ash as a potential adsorbent for the removal of metals from acid mine drainage. Appl Water Sci. 2013 Sep;3(3):567–576.

[30] Feng D, van Deventer JSJ, Aldrich C. Removal of pollutants from acid mine wastewater using metallurgical by-product slags. Sep Purif Technol [Internet]. 2004 Nov;40 (1):61–67. Available from: https://

linkinghub.elsevier.com/retrieve/pii/S1383586604000061

[31] Name T, Engineering CS-M, 2014 undefined. Remediation of acid mine drainage using metallurgical slags. Elsevier [Internet]. [cited 2021 Jun 24]; Available from: https://www.sciencedirect.com/science/article/pii/S0892687514001046

[32] Mackie A, research MW-W, 2012 undefined. Bench-scale study of active mine water treatment using cement kiln dust (CKD) as a neutralization agent. Elsevier [Internet]. [cited 2021 Jun 24]; Available from: https://www.sciencedirect.com/science/article/pii/S0043135411006282

[33] Chockalingam E, Chemosphere SS-, 2006 undefined. Studies on removal of metal ions and sulphate reduction using rice husk and Desulfotomaculum nigrificans with reference to remediation of acid mine. Elsevier [Internet]. [cited 2021 Jun 24]; Available from: https://www.sciencedirect.com/science/article/pii/S0045653505007162

[34] Bogush AA, Voronin VG. Application of a peat-humic agent for treatment of acid mine drainage. Mine Water Environ. 2011 Sep;30(3): 185–190.

[35] El-Awady MH, Sami TM. Removal of heavy metals by cement kiln dust. Bull Environ Contam Toxicol. 1997 Oct; 59(4):603–610.

[36] Erol M, Küçükbayrak S, Ersoy-Meriçboyu A, Ulubaş T. Removal of Cu2+ and Pb2+ in aqueous solutions by fly ash. Energy Convers Manag. 2005 May;46(7–8):1319–1331.

[37] Mackie AL, Walsh ME. Bench-scale study of active mine water treatment using cement kiln dust (CKD) as a neutralization agent. Water Res. 2012 Feb;46(2):327–334.

[38] Muliwa AM, Leswifi TY, Onyango MS. Performance evaluation of eggshell waste material for remediation of acid mine drainage from coal dump leachate. Miner Eng. 2018 Jun;122:241–250.

[39] Ahmad R, Kumar R, Haseeb S. Adsorption of Cu2+ from aqueous solution onto iron oxide coated eggshell powder: Evaluation of equilibrium, isotherms, kinetics, and regeneration capacity. Arab J Chem. 2012 Jul;5(3): 353–359.

Chapter 10

Removal of Copper and Lead from Water in The Mariana Mining Disaster Using Biomass Banana Peel and Rice Husk

Mayra Santos, Vitor Silva, Thainá Andrade, Tamise Aquino, Bruno Batista, Bruna Ferreira, João Mesquita, Márcia Faria, Luiz Maia and Jairo Lisboa Rodrigues

Abstract

The dumping of the mining tailings dam from Mariana, Brazil released about 34 million mining tailings in the Doce river basin, containing many toxic metals. The biomasses of banana and rice were used as adsorbents in the removal of Cu (II) and Pb (II) metals from contaminated water. Quantification of metals was performed using NexION 300D PerkinElmer (USA) ICP-MS. The pH effect studies indicated that the adsorption analyzed in the present work did not undergo significant changes with the variation of the pH values, thus for both banana and rice the best adsorption capacity of Cu (II), 34.11 mg g^{-1} and 34.37 mg g^{-1}, was at pH 5. For Pb (II), the highest adsorption capacity was also at pH 5 with 36.06 mg g^{-1} for banana and 36.04 mg g^{-1} for rice. There was a rapid adsorption where, in all cases in the first 30 minutes of adsorption, more than 60% of the metals had already been adsorbed. Finally, tests were carried out using real samples from Doce river contaminated by the metals under study due to the Mariana disaster. The biomasses presented excellent performance in Cu (II) and Pb (II) removal, reaching concentrations close to zero after adsorption process.

Keywords: adsorption, metals, mining activity

1. Introduction

Pollution from toxic metals is a serious environmental problem all over the world. The intense technological and industrial development led to the excessive use of toxic metal ions and, thus, increased water pollution. These metals are often reported to be harmful to humans and other organisms because of their high toxicity and cumulative effects [1].

On November 5, 2015, Brazil faced its worst environmental disaster after an iron-ore mining waste dam collapsed releasing 34 million m^3 of contaminated mud in a headwater region of the Doce River basin [2].

In a report from the Minas Gerais Water Management Institute [3], which monitored some elements in water samples from points in the Doce river Basin, high concentrations of elements such as Pb and Cu, well above maximum value recommended by environmental legislation [4]. In this sense, it is necessary to develop technologies that can remove these pollutants, so that this water is properly used for human and animal consumption.

There are several methods to remove toxic metal ions from aqueous solutions which consist mainly of chemical, physical and biological techniques [5]. However, these technologies are not always economically viable for Water Treatment Plants.

A promising alternative is the use of biomass adsorption (biosorption). It has low cost, wide adaptation and selectivity in the removal of different types of toxic metals and stable performance in the purification of wastewater (1-100 mg L^{-1}) [6]. The use of plant biomass to obtain raw material is one of the proposals of green chemistry. Green chemistry is an approach that aims to minimize the environmental impact through the creation, development and application of chemical products and processes for reduction or elimination of the use and the generation of toxic substances to the detriment of its treatment [7].

Biosorption is one of the most promising technologies for the remediation of aquatic areas polluted with toxic metal ions [8]. There are several compounds with promising characteristics for the treatment by the use of biosorption, among them rice straw and banana peel stand out. Rice straw contains large amounts of cellulose, hemicellulose, lignin and silica, these compounds provide binding sites for metals [9]. The banana peel has been used as a biosorbent for adsorption of copper, and it is readily available, inexpensive and ecologically correct [10].

The aim of this study was to evaluate and compare the adsorption capacity of the biomass of rice straw and banana peel in waters contaminated by Cu (II) and Pb (II) metals and application in contaminated waters due Mariana disasters.

2. Materials and methods

2.1 Origin of materials

The rice husk originates from the Muriaé region, in the state of Minas Gerais (Brazil) where it is widely grown and easy to obtain biomass. The banana peel was obtained in fairs of the city of Teófilo Otoni – MG (Brazil), these fairs are supplied by small producers of all region of the Mucuri Valley.

2.2 Preparation of biomass

2.2.1 Banana peel

The preparation of the biomass obtained from the banana peel was washed with ultrapure water, taken to the oven for drying at 70° C and ground by a SHOP63 01 industrial blender. After trituration the obtained product was sieved for Bertel granulometric analysis at 40 mesh.

2.2.2 Rice husk

The procedure for obtaining the rice husk biomass resembles the banana peel process. The rice husk was prepared by rinsing the husk with ultrapure water for the removal of particles, and was then taken to an oven at 70°C for 24 hours for drying. After drying, they were crushed in an electric mill and its product passed in a sieve for Bertel granulometric analysis with 40 mesh opening.

2.3 Characterization of materials

2.3.1 Scanning electron microscopy (SEM)

Scanning electron microscopy (SEM) images for biomass samples of banana peel and rice were obtained on a Zeiss (FEG-SEM) Sigma VP model.

2.3.2 Infrared spectroscopy with fourier transform (FT - IR)

IR-FT spectra were obtained on a Perkim Elmer FT-IR / NIR Spectrometer Frontier equipment at a frequency range of 4000 and 500 cm^{-1}. The examples were prepared using the KBr disk method.

2.3.3 Zeta potential

Zeta potential of banana peel biomass and rice straw were measured using a Zetasizer Nano-ZS (Malvern Instruments, UK).

2.4 Solutions of toxic metal ions

Solutions were prepared to be used in the isotherm, pH and kinetics tests. For each metal (Cu (II) and Pb (II)), a 1000 mg L^{-1} solution was used as the standard solution for other solutions of concentration: 0.5, 5, 20, 50, 100 and 200 $mg.L^{-1}$. The contaminating reagents used were Dehydrated Copper (II) Chloride - Sigma 99% and Lead (II) Nitrate - Cinética.

2.5 Adsorption tests

2.5.1 Isotherm test

A quantity of 10 mg of the adsorbent were used in 10 ml samples of the metal solutions at concentrations of 0, 0.5, 5, 20, 50, 100 and 200 $mg.L^{-1}$. Afterwards, they were placed in a Thermo Scientific shaker model 4360 for 24 hours. After 24 hours, the samples were taken to a CIENTEC CT-6000R centrifuge for 3 minutes at 5500 rpm (4058 g) and the aliquot removed to dilute to 2% nitric acid. The initial and residual concentrations were analyzed by NexION 300D PerkinElmer (USA) ICP-MS and the adsorbed amount (q_e) was calculated by Eq. (1):

$$q_e = \frac{V(C_o - C_e)}{m} \quad (1)$$

where C_o and C_e ($mg.L^{-1}$) are the initial and equilibrium concentrations respectively, V is the volume of the solution (L) and m is the mass (g) of the adsorbent used in the experiments.

The Langmuir model (Eq. (2)) is used for a monolayer adsorption process on a homogeneous surface, in which the concentration occurs at specific sites in the adsorbent. The equation of the Langmuir isotherm is given by Eq. (2):

$$q_e = \frac{K_L Q_m C_e}{1 + K_L C_e} \quad (2)$$

where Q_m ($mg.g^{-1}$) and K_L ($L.mg^{-1}$) are the maximum adsorption capacity and the Langmuir constant used for adsorption energy, respectively.

The adsorption is favorable ($0 < R_L < 1$), unfavorable ($R_L > 1$), linear ($R_L = 1$) or irreversible ($R_L = 0$) (Eq. (3)) [11].

$$R_L = \frac{1}{1 + K_L C_o} \tag{3}$$

The Freundlich model (Eq. (4)) is used to describe multilayer adsorption on heterogeneous surfaces.

$$q_e = K_F C_e^{1/n} \tag{4}$$

where K_F and $1/n$ are the Freundlich constants representing adsorption capacity and adsorption intensity, respectively.

2.5.2 pH test

For pH test, 10 mg of the adsorbent were used in 10 ml samples of the metal solutions at a concentration of 20 ppm of the metal solutions at pHs of 5, 7 and 9. With the pH corrected, the solutions were placed in the Thermo Scientific shaker model 4360 for 24 hours. After 24 hours of stirring, the samples were centrifuged in a CIENTEC CT-6000R centrifuge for 3 minutes at 5500 rpm (4058 g) for aliquot withdrawal, for further dilution in 2% nitric acid, and analysis by ICP-MS Nexion 300D PerkinElmer (USA).

2.5.3 Kinetic test

The metallic solution of 10 ml to 20 $mg.L^{-1}$ was added to 10 mg of the adsorbent material (biomass of banana peel and rice). Soon after, the solutions had the pH corrected to 7 and placed in a Thermo Scientific model 4360 agitator, with variations in contact time of the biomass with the metallic solution: 0.5, 2, 4, 6, 12 and 24 hours. After the stirring time, the samples were centrifuged through CIENTEC CT-6000R for 3 minutes at 5500 rpm (4058 g) for aliquot extraction, dilution (2% HNO_3) and analysis of ICP-MS NexION 300D PerkinElmer (USA). The kinetic adsorption data were adjusted with the models of pseudo-first order [12] (Eq. (5)) and pseudo-second order [13] (Eq. (6)) models:

$$q_t = q_e \left(1 - e^{-k_1 t}\right) \tag{5}$$

$$q_t = \frac{k_2 q_e^2 t}{1 + k_2 q_e t} \tag{6}$$

where qe and qt are the amounts of adsorbent ($mg\ g^{-1}$) in equilibrium and at time t (min), respectively. k_1 (min^{-1}) and k_2 ($g\ mg^{-1}\ min^{-1}$) are the equilibrium adsorption rate constants for pseudo-first and pseudo-second order adsorption, respectively.

2.6 Collecting contaminated water by metals in the Doce river basin

The water samples were collected in two different areas of the Doce river basin, in the state of Minas Gerais, in the years 2016, 2017 and 2018. The areas were chosen in relation to the contamination gradient. Two samples were collected at each point (**Figure 1**) whose latitude and longitude are described below: P1- 18° 52′ 55.31 (S) 41° 57′ 86″(W) and P2 - 18° 51′ 22.54″ (S) 41° 56′ 10.55″(W). The first point is located in the water and sewage treatment system and the second point in the city center of Governador Valadares, Brazil.

The water samples were collected according to the procedure adopted by the Environmental Sanitation Technology Company [14]. Briefly, 1000 mL vials of Falcon BD® model polypropylene were contaminated free of contaminants were used for collection. Water samples were stabilized with ultrapure nitric acid (0.5% HNO_3). The concentrations of the metals in the samples were determined using ICP-MS according to EPA method 200.8 [15].

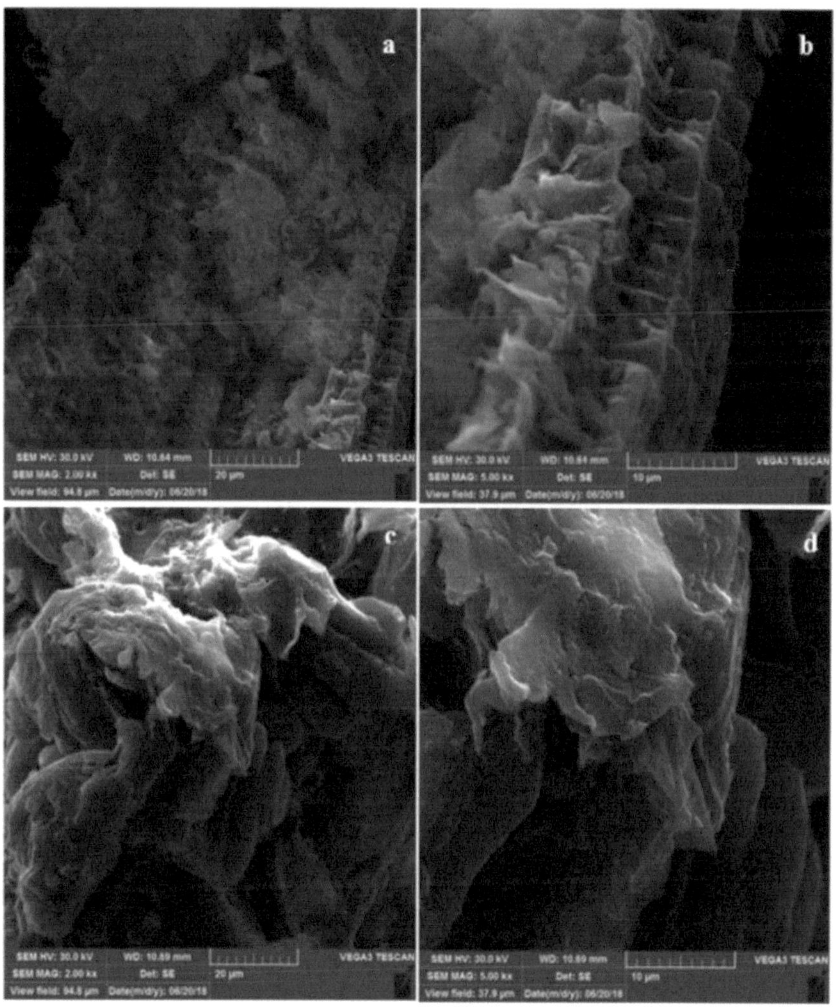

Figure 1.
Scanning electron microscopy image of (a - b) rice (c - d) banana biomass expand 2.000 and 5.000 times.

3. Results and discussion

3.1 Characterization

3.1.1 Scanning electron microscopy

The **Figure 1(a)** and **(b)** shows the scanning electron microscopy of rice biomass, with magnification of 2.000 x in (a) and 5.000 x in (b). In (c) and (d), it shows the banana biomass, with an image enlarged from 2.000 x in (c) to 5.000 x in (d). In all the images it is possible to observe a porous structure, being able to be susceptible to adsorption of metals on its surface.

3.1.2 Infrared spectroscopy with fourier transform

The main adsorption bands were observed in the infrared region (500-4000 cm^{-1}) and compared with the literature.

In **Figure 2**, the banana peel biomass presents: the absorption band of 3200 cm-1 can be attributed to the O-H bond stretch, characteristic of functional groups of alcohols and phenols. Pino [16] as well as the presence of H_2O molecules. The wavenumber 2930 cm^{-1} corresponds to the axial deformation of aliphatic and hydro aromatic carbon sp^3 carbon bonds found in cellulose and hemicellulose [17]. In the vicinity of the range 1620 cm^{-1} indicates the lignin, due to the functional group C=O [18]. Finally, the peak near 1000 cm^{-1} is assigned to the C-O group stretching than can be observed in cellulose, hemicellulose and lignin. Thus, it can be concluded that groups alcohols/phenols, carboxyl, carbonyl, alkane, aromatic groups are present in banana bark [19].

Rice bark biomass shows major bands in the bands of 3369 cm^{-1} indicating O-H elongation, revealing a hydrogen bond, this structure may be due to the presence of acids or alcohols [20]. The wavelength 2900 cm^{-1}, as well as in banana biomass,

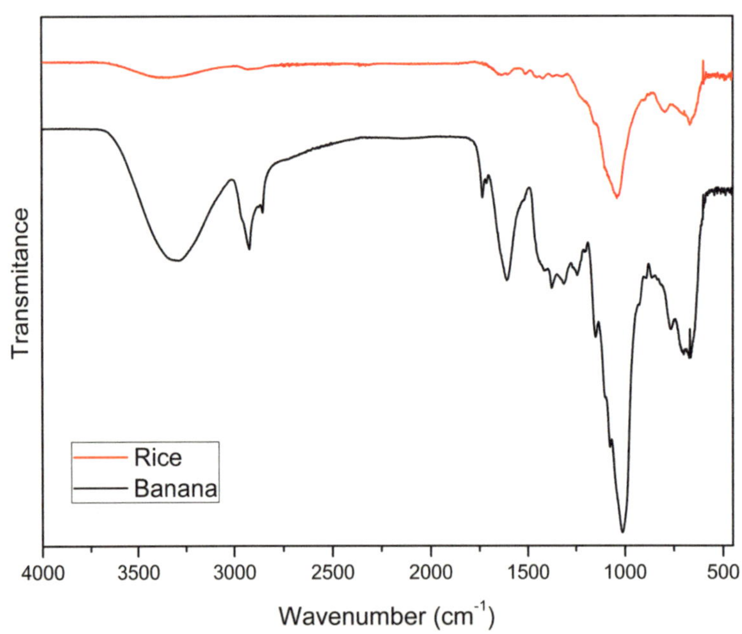

Figure 2.
Results of the Fourier transform infrared analysis of rice and banana biomass.

indicates groups of C-H, from carbon sp^3, connecting to the presence of cellulose and hemicellulose. The spectral band around 1500-1640 cm^{-1} attributes to the functional group carbonyl (C=O) elongation associated with lignin [21]. The bands near 1100 and 800 cm^{-1} are characteristic of the O-Si bond in polymorphic silica, SiO_2 [22].

The adsorption mechanism of the copper and lead ions in the studied biomass can be associated with the presence of functional groups that have oxygen in their constitution [23]: a two-step process: (i) the metal ions hydrolyze in the solution to form a hydrolyzed metal ion (ii) the positively charged halves interact with oxygen-containing functional groups, leading to better contact with biomass and high adsorption capacity.

3.1.3 Zeta potential

Figure 3 show the zeta potentials of the banana and rice peel. The original peels exhibit a negative zeta potential and the used particles possess a less negative potential. All these results reveal that the waste peels had adsorbed the metal ions. The isoelectric point for banana was 6.6 – 7.0 and for rice 6.6 – 7.3.

3.2 Adsorption tests

3.2.1 Isotherm test

3.2.1.1 Copper

In order to determine and compare the adsorption capacity of copper by the banana and rice biomass, **Figure 4** shows two graphs of the Cu (II) equilibrium concentration (C_e) as a function of the adsorption capacity (Q_e) of the biomasses. **Figure 4(a)** shows a marked increase in the isotherm, indicating that the free sites of the banana biomass were empty and available for Cu (II) adsorption. While **Figure 4(b)** shows some initial oscillation in the adsorption capacity but soon after the free sites reach the saturation and, consequently, a balance in the adsorption capacity of the rice biomass.

The Langmuir and Freundlich isotherm models were tested for information on Cu (II) adsorption by banana and rice biomasses. These models describe the interaction between the adsorbent and the adsorbed material. The Langmuir model considers that the adsorption process occurs in a monolayer on a homogeneous

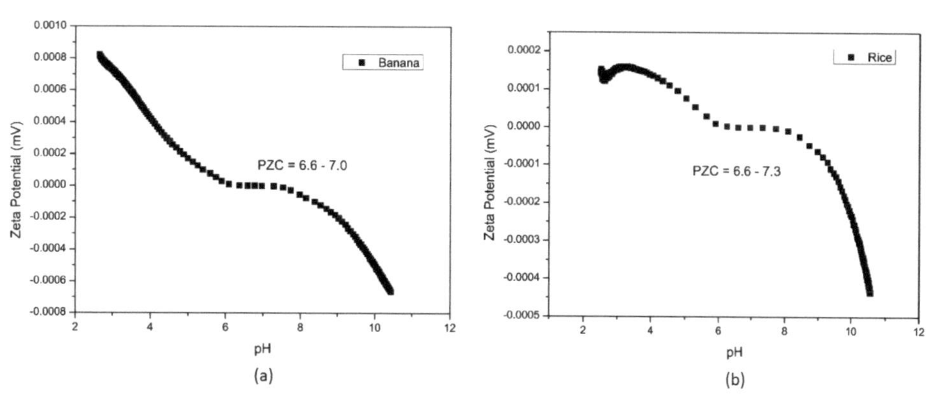

Figure 3.
Zeta potential curves obtained for (a) banana peel and (b) rice rusk.

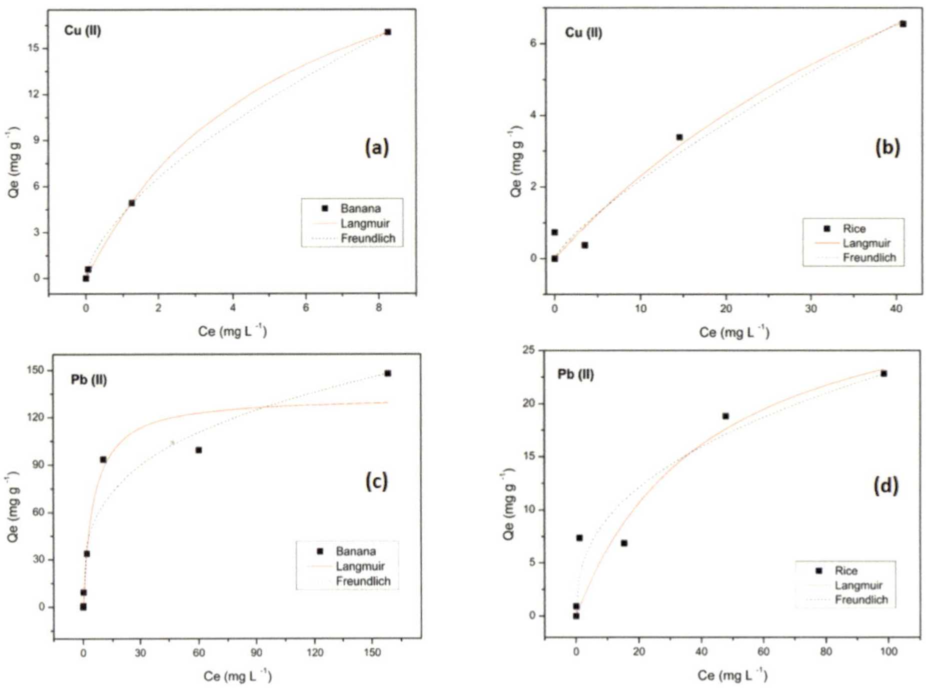

Figure 4.
Adsorption isotherm of Cu (II) (a) by banana, (b) by rice and of Pb (II) (c) by banana, (d) by rice. Adsorbent dose: 10 mg; shaking speed:200 rpm; pH 7 ± 0.3; temperature: 25 ± 0.3°C.

surface, that is to say, with active sites with identical energy and availability and without interaction between the adsorbed molecules [24].

The Freundlich isotherm differs from the Langmuir isotherm model, which describes multilayer adsorption on heterogeneous surfaces [25]. In the Freundlich model, when the 1/n exponent value for this model equals 1, the adsorption is considered linear, with identical adsorption energies at all sites. The higher the value of 1/n, the stronger the interaction between adsorbent and adsorbate [26].

Table 1 shows the values of the parameters obtained by nonlinear regression of the Cu (II) adsorption by the banana and rice biomass applied to the Langmuir and Freundlich isotherm. For banana biomass, the Langmuir and Freundlich isotherms presented very close adaptations when comparing the values of the correlation coefficient (R^2): 0.99 (Langmuir) and 0.99 (Freundlich). In addition, the R_L value is between 0 and 1, indicating that the Langmuir model is favorable, but Freundlich is also an accepted model for this adsorption since the value of 1/n is less than 1. For rice biomass, the Langmuir model was more favorable, with $R^2 = 0.96$ and $R_L = 0.99$.

3.2.1.2 Lead

The **Figure 4(c)** and **(d)** illustrates the experimental data obtained in the adsorption of Pb (II) by banana and rice biomasses applied to the Langmuir and Freundlich models. For both banana and rice biomass, there was an increase in the adsorption capacity, indicating the occupation of the free sites in the material.

The adsorption of Pb (II) by banana biomass was favorable to the Langmuir isotherm model with $R^2 = 0.95$ and $R_L = 0.92$. For rice biomass the Freundlich model presented $R^2 = 0.91$, but another parameter, 1/n, indicates that this model is not adequate because it has a value higher than 1. With this, the Langmuir model is the most favorable model.

Isotherm	Parameters	Biomass	
		Banana	Rice
Cu (II)			
Langmuir	Q_m (mg g^{-1})	26.68	16.80
	K_L (L mg^{-1})	0.18	0.02
	R^2	0.99	0.96
	R_L	0.92	0.99
Freundlich	K_F (L g^{-1})	4.23	0.35
	1/n	0.63	0.79
	R^2	0.99	0.95
Pb (II)			
Langmuir	Q_m (mg g^{-1})	133.71	33.33
	K_L (L mg^{-1})	0.18	0.02
	R^2	0.95	0.86
	R_L	0.92	0.99
Freundlich	K_F (L g^{-1})	32.21	3.67
	1/n	0.30	2.52
	R^2	0.93	0.91

Table 1.
Parameters obtained by nonlinear regression adjustment of Langmuir, Freundlich, for Cu (II) and Pb (II) isotherms by adsorption in biomass of banana and rice husk. Q_m: Maximum adsorption capacity, K_L: Langmuir constant, R_L: Balance parameter, K_F: Freundlich constant, 1/n: Adsorption intensity.

3.3 pH test

3.3.1 Copper

The effect of pH on copper (II) adsorption by banana and rice biomass at pH values of 5.0, 7.0 and 9.0 is described in **Figure 5(a)**. The adsorption capacity of the metal by the banana is higher at pH 5.0 (34.16 mg g^{-1}) than pH 7.0 (33.80 mg g^{-1}) and 9.0 (31.02 mg g^{-1}). In agreement, the rice presents similar behavior where the capacity of adsorption undergoes a slight decrease with the increase of the pH, pH 5.0 (34.37 mg g^{-1}), pH 7.0 (33.70 mg g^{-1}) and pH 9.0 (30.81 mg g^{-1}). As in the literature [27], the rate of Cu (II) removal has little effect on pH.

3.3.2 Lead

Figure 5(b) shows the effect of pH on the adsorption of Pb (II) by banana and rice biomasses. Banana biomass presented a removal rate of 91.21% (36.46 mg g^{-1}) at pH 5, 93.88% (37.53 mg g^{-1}) at pH 7.0 and 94.46% (37.76 mg g^{-1}) to pH 9.0. As for banana, the adsorption capacity for rice did not have a significant effect, since at pH 5.0 it removed 93.61% (37.42 mg g^{-1}) from Pb (II) 94.85% (37. 91 mg g^{-1}) and 95.68% (38.25 mg g^{-1}) to pH 7.0 and 9.0, respectively.

The behavior shown where the adsorption capacity has a small increase together with the pH value is justified by the PZC values of banana biomass (6.6 - 7.0) and rice (6.6 - 7.3) and a study [28] where it states that the negative surface charge leads to deprotonation of functional groups of the biomass as H^+ (aq) and H_3O^+ (aq) are

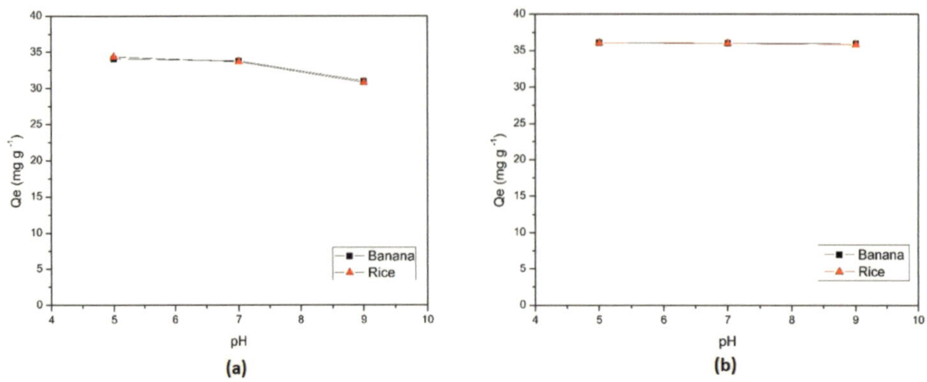

Figure 5.
Effect of pH on (a) Cu (II) and (b) Pb (II) adsorption by banana and rice biomass. Initial [Cu^{2+}] and [Pb^{2+}]: 20 mg L^{-1}; adsorbent dose: 20 mg; shaking speed: 200 rpm; temperature: 25 ± 0.3°C.

released from them. Thus, these deprotonated functional groups serving as binding sites become readily accessible to metal ions causing better sorption.

3.4 Kinetic test

3.4.1 Copper

The results obtained from the kinetic copper adsorption test for banana and rice biomass are presented in **Figure 6(a)** and **(b)**. At pH 7.0, the results indicate a rapid removal in the first 30 minutes of contact, with approximately 53.16% for banana and 69.66% for rice. The maximum removal rate was reached at 120 minutes for banana, with 64.73% and 360 minutes for rice, with 91.72%.

The kinetic adsorption data were fitted with a pseudo first order model [12] and pseudo second order model [13].

The kinetic pseudo-first order model does not adjust to copper adsorption by banana and rice biomasses. While the pseudo-second order model based on the adsorption capacity of the solid phase shows the processes of adsorption studies in all time bands.

The kinetic data presented linearity (**Figure 6(b)**), with a correlation coefficient of 0.99 for banana and 0.99 for rice. The qe calculated with values close to experimental qe. For banana, the values qe (14.05 mg g^{-1}), K_2 (0.0012 L mg^{-1}) and qe (29.41 mg g^{-1}), K_2 (0.0026 L mg^{-1}) were obtained from the slope and intersection of the straight line of the graph t/qt as a function of t, according to **Figure 6(b)**.

3.4.2 Lead

Figure 6(c) and **(d)** shows the results obtained for the kinetic test at the times of 30, 120, 360, 720 and 1440 minutes. At pH 7.0, the data indicate a rapid adsorption of lead by banana biomass in the first 30 minutes of contact, with approximately 83% removal. While the biomass of rice removed about 75.9% in the first 120 minutes.

The kinetic adsorption data were fitted with a pseudo first order model and pseudo second order model. The pseudo-first-order kinetic model does not fit the adsorption of lead by banana or rice biomass. While the pseudo-second order model

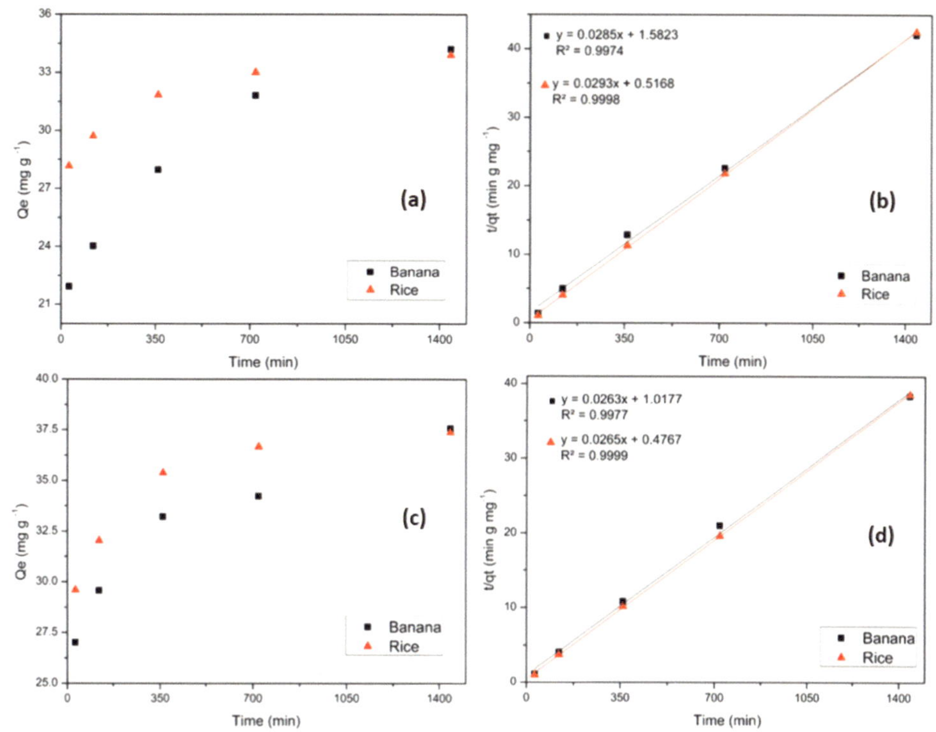

Figure 6.
(a) Kinetics of Cu (II) adsorption by banana and rice biomass (b) pseudo-second order kinetic model of Cu (II) adsorption by banana and rice biomass (c) kinetics of Pb (II) adsorption by banana and rice biomass (d) pseudo-second order kinetic model of Pb (II) adsorption by banana and rice biomass. Initial [Cu^{2+}] and [Pb^{2+}]: 20 mg L^{-1}; pH: 7.0 ± 0.3; shaking speed: 200 rpm; temperature: 25.0 ± 0.3°C.

based on the adsorption capacity of the solid phase shows the adsorption process in all time bands.

The kinetic data for the adsorption of lead by banana biomass present a high linearity (**Figure 6(d)**), with correlation coefficient close to 1 (0.99) and calculated Q_e (26.60 mg g^{-1}) with values close to experimental Q_e (26.96 mg g^{-1}). The values of calculated Q_e and K_2 (0.0014 L mg^{-1}) were obtained from the slope and intersection of the straight line of the graph t/qt as a function of t.

For the rice biomass, the correlation coefficient is also close to 1 (0.99), the calculated Q_e (23.81 mg g^{-1}) presented a value close to the experimental Q_e (23.92 mg g^{-1}) at the end the K_2, through the 0.0006 L mg^{-1} slope of the line.

The adsorption capacities of copper and lead in banana and rice biomass are compared with the studies on the types of materials of organic origin (**Table 2**). In addition to the efficiency of the removal, it is easy to obtain and handle the adsorption capacity. Considering this, it is noticed that the studied materials present great efficiency in the removal of metals from contaminated water. Worked biomasses are cheap and easy to prepare because they are reused materials from common productive activities in various regions, making this technology accessible to various social levels.

Comparing the biomaterials of this study with the values found in the literature (**Table 2**), the biomasses of banana peel and rice present better adsorption capacities, thus confirming the great potential of applying the material with real water samples.

Adsorbent	Adsorption Capacity ($mg.g^{-1}$)	pH	Reference
Cu (II)			
Aspergillus niger-treated rice straw	23.80	4.0 – 6.0	[29]
Pine sawdust	9.59	7.0	[30]
Treated laterite	7.25	5.0	[31]
Hazelnut husk	6.64	5.7	[32]
Irish peat moss	17.60	4.5	[33]
Banana peel	26.68	7.0	Present study
Rice husk	16.80	7.0	Present study
Pb (II)			
Organo-mineral complex	1.59	7.0	[34]
Cucumber peel	133.60	5.0	[28]
Palm biomass	98.90	5.5	[35]
F. andina fibers	31.50	7.0	[36]
Biofilm on F. andina	35.90	7.0	[36]
Banana peel	133.71	7.0	Present study
Rice husk	33.33	7.0	Present study

Table 2.
Comparison between the adsorption capacities of different materials in copper and lead removal.

3.5 Application of the materials in real samples from environmental mariana disaster region

The **Figure 7** shows the concentration of metals (a) Cu (II) and (b) Pb (II) before and after adsorption by using banana and rice biomasses at two points collected in the city of Governador Valadares - MG in the years of 2016, 2017 and 2018. For copper, the concentration at point 2 is higher when compared to point 1 in the

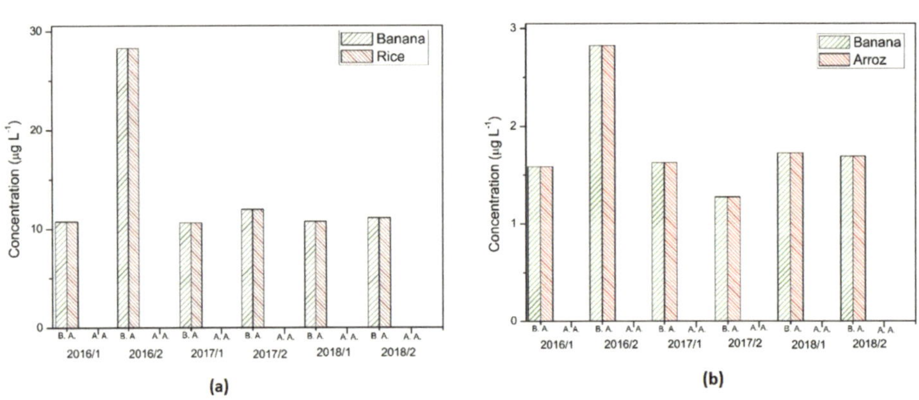

Figure 7.
(a) Adsorption of Cu (II) onto banana and rice biomass in samples collected from two points of a river in the region of Governador Valadares, state of Minas Gerais, Brazil, (b) adsorption of Pb (II) onto banana and rice biomass in samples collected from two points of a river in the region of Governador Valadares, state of Minas Gerais, Brazil, (B.A- before adsorption; a.A- after adsorption; 2016/1- sample collected in 2016 at point 1; 2016/2- sample collected in 2016 at point 2; 2017/1- sample collected in 2017 at point 1; 2017/2- sample collected in 2017 at point 2; 2018/1- sample collected in 2018 at point 1; 2018/2- sample collected in 2018 at point 2).

same year, 2016: p1 (10.75 μg.L^{-1}) and p2 (28.29 μg.L^{-1}), 2017: p1 (10.63 μg.L^{-1}) and p2 (11.98 μg.L^{-1}), 2018: p1 (10.75 μg.L^{-1}) and p2 (11.13 μg.L^{-1}). For lead, with the exception of the year 2016, p1 (1.58 μg.L^{-1}) and p2 (2.82 μg.L^{-1}), the same trial was observed, 2017: p1 (1.62 μg.L^{-1}) and p2 (1.27 μg.L^{-1}), 2018: p1 (1.72 μg.L^{-1}) and p2 (1.6911 μg.L^{-1}).

Point 2, in all years, presented higher concentrations when compared to point 1, for both Cu (II) (28.29 μg.L^{-1}) and Pb (II) (2.82 μg.L^{-1}). According to the US EPA, the maximum limits for copper and lead are 1300.00 μg L^{-1} and 15.00 μg.L^{-1}, respectively. In spite of this, the concentrations of the studied metals reached very close to zero after the treatment with banana and rice biomasses.

Seen this, the studied material shown efficiency in removal of copper and lead in contaminated water even in real samples with several parameters outside the ideal that cannot be controlled and altered. With this, these biomaterials have application in real situations of contamination by metals with high removal rate and low cost [37].

4. Conclusion

Finally, it can be concluded that the studies presented achieved results beyond expectations, with high adsorption capacity. Isotherm studies demonstrate that the Langmuir model is the most applicable in all analyzed cases. Studies of the effect of pH indicated that the adsorption analyzed in the present work did not undergo significant changes with the variation of pH values. Where, for both banana and rice, the best Cu (II) adsorption capacity, 34.11 mg.g^{-1} and 34.37 mg.g^{-1}, respectively, was at pH 5. For Pb (II), the highest adsorption capacity was also at pH 5 with 36.06 mg.g^{-1} for bananas and 36.04 mg.g^{-1} for rice. In the kinetics studies, there was a fast adsorption, in all cases, in the first 30 minutes of adsorption more than 60% of the metals had already been adsorbed. For tests on real samples, the biomasses showed excellent performance in removing Cu (II) and Pb (II), reaching concentrations close to zero after adsorption, indicating efficiency in environmental remediation. In view of this, the biomasses studied can be used to purify water contaminated by mining or other sources. For future studies, the application of biomass in the form of filters will be analyzed.

Acknowledgements

The present study was developed in the scope of a post-graduation, master's dissertation entitled: "Use of anana skin biomass and rice skin in removal copper (II) and lead (II) of water samples from the region of Mariana environmental disaster", at the Federal University of the Valleys of Jequitinhonha and Mucuri.

Author details

Mayra Santos[1], Vitor Silva[1], Thainá Andrade[1], Tamise Aquino[1], Bruno Batista[2], Bruna Ferreira[2], João Mesquita[3], Márcia Faria[1], Luiz Maia[1] and Jairo Lisboa Rodrigues[1*]

1 Federal University of the Jequitinhonha and Mucuri Valleys, Teófilo Otoni, Brazil

2 Federal University of ABC, Santo André, Brazil

3 Federal University of The Jequitinhonha and Mucuri Valleys, Diamantina, MG, Brazil

*Address all correspondence to: jairo.rodrigues@ufvjm.edu.br

References

[1] Deniz F, Karabulut A, Biosorption of heavy metal ions by chemically modified biomass of coastal seaweed community : Studies on phycoremediation system modeling and design. Ecol. Eng. 2017; 106: 101-108. https://doi.org/10.1016/j.ecoleng.2017.05.024

[2] Rudorff N, Rudorff C M, Kampel M, Ortiz G, Sensing monitoring of the impact of a major mining wastewater disaster on the turbidity of the Doce River plume off the eastern Brazilian coast. ISPRS Journal of Photogrammetry and Remote Sensing Remote. 2018. https://doi.org/10.1016/j.isprsjprs.2018.02.013

[3] Instituto Mineiro de Gestão das Águas (IGAM), Monitoramento da Qualidade das Águas Superficiais do Rio Doce no Estado de Minas Gerais. 75p. 2016. http://www.igam.mg.gov.br/.

[4] Conselho Nacional do Meio Ambiente (CONAMA), Resolução n° 357/2005, de 17 de março de 2005. Dispõe sobre a classificação dos corpos de água e diretrizes ambientais para o seu enquadramento, bem como estabelece as condições e padrões de lançamento de efluentes, e dá outras providências. Brasília, DF: 2005. http://www2.mma.gov.br/port/conama/legiabre.cfm?codlegi=459.

[5] Kumar R, Mudhoo A, Lofrano G, Chandra M, Biomass-derived biosorbents for metal ions sequestration : Adsorbent modification and activation methods and adsorbent regeneration. Journal of Environmental Chemical Engineering. 2014; 2: 239-259. https://doi.org/10.1016/j.jece.2013.12.019

[6] Sud D, Mahajan G, Kaur M P, Agricultural waste material as potential adsorbent for sequestering heavy metal ions from aqueous solutions – A review. Bioresource Technology. 2008;99: 6017-6027. ttps://doi.org/10.1016/j.biortech.2007.11.064

[7] Lenardão E J, Dabdoub M J, Batista C F, "Green chemistry" – os 12 princípios da química verde e sua inserção nas atividades de ensino e pesquisa. Química Nova.2003; 26: 123-129. http://dx.doi.org/10.1590/S0100-40422003000100020

[8] Gupta V K, Nayak A, Agarwal S, Bioadsorbents for remediation of heavy metals : Current status and their future prospects. Environ. Eng. Res. 2015; 20: 1-18. http://dx.doi.org/10.4491/eer.2015.018.

[9] Gao H, Liu Y, Zeng G, Xu W, Li T, Xia W, Characterization of Cr (VI) removal from aqueous solutions by a surplus agricultural waste-Rice straw. J. Hazard. Mater. 2008; 150: 446-452. https://doi.org/10.1016/j.jhazmat.2007.04.126.

[10] Hossain M A, Ngo H H, Guo W S, Nguyen T V, Biosorption of Cu (II) From Water by Banana Peel Based Biosorbent : Experiments and Models of Adsorption and Desorption. Journal of Water Sustainability. 2012; 2: 87-104.

[11] Weber T W, Chakrovorti R K, Pore and Solid Diffusion Models for 20. 1974; 20: 228-238. https://doi.org/10.1002/aic.690200204.

[12] Lagergren S, About the theory of so-called adsorption of soluble substance. Kungliga Svenska Vetenskapsakademiens Handlingar. 1898; 24:1-39.

[13] Ho Y S, Mckay G, Pseudo-second order model for sorption processes. Process Biochemistry 1999; 34: 451-465.

[14] Companhia Ambiental do Estado de São Paulo (CETESB), 2011, Guia nacional de coleta e preservação de

amostras: agua, sedimento, comunidades aquaticas e efluentes líquidos. http://www.cetesb.sp.gov.br.

[15] United States Environmental Protection Agency (USEPA), 2000, Technologies and costs for removal of arsenic from drinking water. Target. Anal. Branch Stand. Risk Manag. Div. Off. Gr. Water Drink. Water United States Environ. Prot. Agency Washington, D.C. 30:1497-1505. https://www.epa.gov/.

[16] Pino G A H, Biossorção de Metais Pesados Utilizando Pó da Casca de Coco Verde (*Cocos nucifera*). Dissertação (Mestrado). Pontifícia Universidade Católica, Rio de Janeiro, 2005.

[17] Saikia R, Chutia R S, Kataki R, Pant K K, Perennial grass (arundo donax l.) as a feedstock for thermo-chemical conversion to energy and materials. Bioresour. Technol. 2015; 188: 265-272. https://doi.org/10.1016/j.biortech.2015.01.089

[18] Boniolo M R, Biossorção de urânio nas cascas de banana. Dissertação (mstrado). Universidade de São Paulo. São Paulo, 2008.

[19] Salim R M, Jalal A, Chowdhury K, Rayathulhan R, Yunus K, Sarkar Z I, Biosorption of Pb and Cu from aqueous solution using banana peel powder 3994. Desalination and Water Treatment. 2015; 1: 1-1:12. https://doi.org/10.1080/19443994.2015.1091613.

[20] Rafiq M K, Bachmann RT, Shang Z, Rafiq M T, Joseph S, Long R, Influence of Pyrolysis Temperature on Physico-Chemical Properties of Corn Stover (Zea mays L.) Biochar and Feasibility for Carbon Capture and Energy Balance. PLoS One 11. 2016; 1: 1-17. https://doi.org/10.1371/journal.pone.0156894.

[21] Yang J, Zeng J, Wen L, Zhu H, Jiang Y, John A, Yu L, Yang B, Effect of morin on the degradation of water-soluble polysaccharides in banana during softening. Food Chem. 2019; 287: 346-353. https://doi.org/10.1016/j.foodchem.2019.02.100.

[22] Chaves T F, Queiroz Z F, Sousa D N R, Girão J H, Uso da cinza da casca do arroz (CCA) obtida da geração de energia térmica como adsorvente de Zn (II) em soluções aquosas. **Química Nova**, 2009; 32: 1378-1383.

[23] Sun J, Li M, Zhang Z, Guo J, Unravelling the adsorption disparity mechanism of heavy-metal ions on the biomass-derived hierarchically porous carbon. Applied Surface Science. 2019; 471: 615-620. https://doi.org/10.1016/j.apsusc.2018.12.050.

[24] Sepúlveda L A, Cuevas F A, Contreras E G, 2015. Valorization of agricultural wastes as dye adsorbents: Characterization and adsorption isotherms. Environ. Technol. 2015; 36: 1913-1923. https://doi.org/10.1080/09593330.2015.1016119

[25] Rahbar N, Jahangiri A, Boumi S, Khodayar M J, Mercury removal from aqueous solutions with chitosan-coated magnetite nanoparticles optimized using the box-behnken design. Jundishapur J. Nat. Pharm. Prod. 2014; 9: 1-7. https://doi.org/10.17795/jjnpp-15913.

[26] Hott R C, Andrade T G, Santos M S, Lima A C F , Faria M C S, Bomfeti C A, Barbosa F, Maia L F O, Oliveira L C A, Pereira M C, Rodrigues J L, Adsorption of arsenic from water and its recovery as a highly active photocatalyst. Environ. Sci. Pollut. Res. 2016; 23: 21969-21979. https://doi.org/10.1007/s11356-016-7441-3

[27] Salim R M, Chowdhury A J K, Rayathulhan R, Yunus K, Sarkar M Z I, Biosorption of Pb and Cu from aqueous solution using banana peel powder. Desalination and Water Treatment. 2015; 1: 1-1:12. DOI: 10.1080/19443994.2015.1091613.

[28] Basu M, Guha A K, Ray L, Adsorption of Lead on Cucumber Peel. Journal of Cleaner Production. 2017; 151: 603 – 615. http://dx.doi.org/10.1016/j.jclepro.2017.03.028.

[29] Wang J, Cui H, Cui C, Xing D, 2016. Biosorption of copper (II) from aqueous solutions by Aspergillus niger -treated rice straw. Ecol. Eng. 2016; 95: 793-799. https://doi.org/10.1016/j.ecoleng.2016.07.019

[30] Semerjian L, Environmental Technology & Innovation Removal of heavy metals (Cu, Pb) from aqueous solutions using pine (Pinus halepensis) sawdust : Equilibrium , kinetic , and thermodynamic studies. Environ. Technol. Innov. 2018; 12: 91-103. https://doi.org/10.1016/j.eti.2018.08.005

[31] Rani K S, Srinivas B, Naidu K G, Ramesh K V, 2018. Removal of copper by adsorption on treated laterite. Materials Today: Proceedings . 2018; 5: 463-469. https://doi.org/10.1016/j.matpr.2017.11.106

[32] Imamoglu M, Tekir O, Removal of copper (II) and lead (II) ions from aqueous solutions by adsorption on activated carbon from a new precursor hazelnut husks. Desalination. 2008; 228: 108-113. DOI:10.1016/j.desal.2007.08.011

[33] Gupta B S, Curran M, Hasan S, Ghosh T K, 2009. Adsorption characteristics of Cu and Ni on Irish peat moss. Journal of Environmental Management. 2009; 90: 954-960. https://doi.org/10.1016/j.jenvman.2008.02.012

[34] Fan C, Du B, Zhang Y, Ding S, Gao Y, Chang M, Adsorption of lead on organo-mineral complexes isolated from loess in Northwestern China. Journal of Geochemical Exploration. 2016; 1: 1-7. https://doi.org/10.1016/j.gexplo.2016.02.012

[35] Amin M T, Alazba A A, Shafiq M, Application of biochar derived from date palm biomass for removal of lead and copper ions in a batch reactor: kinetics and isotherm scrutiny. *Chemical Physics Letter. 2019;* https://doi.org/10.1016/j.cplett.2019.02.018

[36] Rodríguez J J G, Rivera A C R, Bennevitz M R V, Living biomass supported on a natural-fiber biofilter for lead removal. Journal of Environmental Management 231 (2019) 825-832. https://doi.org/10.1016/j.jenvman.2018.11.004

[37] Santos M S, Uso de biomassa de casca de banana e casca de arroz na remoção de cobre (II) e chumbo (II) de amostras de água da região do desastre ambiental de Mariana, Dissertação (mestrado), Universidade Federal dos Vales do Jequitinhonha e Mucuri, Teófilo Otoni - MG, 2019.